# Personal Development Guide

The COA guide to gaining work experience, exploring gap year activities, improving your grades and getting a taste of higher education

Edited by
Guy Nobes

COA
Clear and focused thinking

# Personal Development Guide

This 12th edition published in 2018 by Cambridge Occupational Analysts Ltd

Editorial and Publishing Team

Editor Guy Nobes
Cover Design PFD
Design and typesetting Simon Foster and Paul Rankin
Illustrations Diana Mainstone



British Library Cataloguing in Publication Data
A catalogue record for this book is available from the British Library.

ISBN 978-1-906711-32-0



Typeset by Cambridge Occupational Analysts Ltd
Printed and bound in the EU by Graphius, Ghent, Belgium

# Contents

# Introduction

This book is about exploring ideas that could help you in your future personal and career development. It is aimed primarily at young people aged around 16 to 18. We envisage that most readers will be working towards AS/A levels, IB, Highers or equivalent and will be in the sixth form/Year 12 or Scottish Year 5 (Y12/S5) at school or college. You may be starting to research possible courses in higher education, perhaps contemplating a Gap Year before going to university, and wondering what on earth you'll be able to write in your application that will have admissions tutors queuing up to recruit such an admirable student!

Our goal is to present you with a series of suggestions and exercises that will give you an idea of the many opportunities available and an understanding of how best to match these opportunities to your own needs.

In the first section of the book, we offer our own broad classification of the astonishing range of placements, courses, projects, trips and treks that you could use to nurture your own set of personal, learning and thinking skills.

Our four headings are:

1. Work Experience
2. Community, Conservation and Heritage Projects in the UK
3. International Community and Conservation Projects, Treks and Expeditions
4. Academic, Cultural, Scientific, Technological, Teaching and Sports-related Options

Should you wish to improve your grades by re-taking examinations or by seeking extra tuition for your current studies, the third section of our book lists colleges offering this type of provision. Finally, we give details of Taster courses, provided by many universities and colleges to allow you to experience something of the academic and social life of an undergraduate student before you make a formal application for a place in higher education.

## Developing essential skills

Whatever you decide to do, try to look beyond the admittedly important elements of fun and excitement to think about whether your chosen activities will help develop the skills regarded as essential for success in education, training, work and life in general.

These skills involve written and spoken communication, using numbers and using computers. They are relevant for everyone, from the youngest pupils in school to the most senior executives in global organisations. With the help of this book, not to mention time spent analysing your needs and exploring appropriate options, you should be able to improve the quality of your learning and performance by acquiring the ability to apply essential skills in different contexts.

You can spend your time in the UK or overseas; you may be paid or you may have to pay a considerable sum in order to participate; you may choose a course of study or you may work with vulnerable children in a threatened environment; or maybe you have simply been stuck in the classroom for too long and need to get out in the fresh air and discover the delights of the great outdoors. This could feature environmental conservation work, sports coaching or some madcap white-knuckle adventure. The choice is yours.

# Personal Development Opportunities

# Work Experience

Work experience is a very broad term that can be interpreted in many different ways. For the purposes of this Personal Development Guide, we see school-based work experience as an opportunity for you to spend time on an employer's premises, observing the workplace, undertaking tasks (under supervision) alongside people working there, and learning about the skills required and behaviour expected.

In this sense, work experience plays a significant role in bridging the gap between education and the world of work. It can help you to:

- become aware of jobs you have not previously considered
- be better informed about career choices
- develop relevant occupational and behavioural skills

School-based work experience is most effective when it is organised as an integral part of your study programme. You should ideally be able to discuss your placement in advance with the employer and with a work experience coordinator in your school to plan a genuine learning experience suited to your needs. The programme should indicate clear roles, responsibilities and expectations both for you and for the employer. Many people find it useful to keep a diary during the placement and to have a debriefing session afterwards to establish what has been learned. This should include feedback from the employer based on your performance.

It should be clear that work experience at this stage is not necessarily about 'trying out' a career in which you might be interested, nor is it the same as a part-time job – in the evening, at weekends or during the holidays – where the emphasis is primarily on earning money rather than on what you can learn from being in a work environment. In fact, you would not normally expect to be paid for a work experience placement, although your employer may consider providing support to cover travel expenses or meal costs.

If your current study programme is academic, you may feel that work experience is not immediately relevant. It is, however, a valuable enrichment activity, which can help you reflect on your progress to date and prepare for the next stage in your personal development. This may be further or higher education, a gap year or employment with training such as an apprenticeship.

Why work experience matters

- Relevant work experience is rated by 66% of recruiting employers as being a critical or significant factor looked for in candidates
(UK Commission for Employment and Skills Employer Perspectives Survey 2014)
- Over a third of recruiters who took part in the research repeated their warnings from previous years – that graduates who have had no previous work experience at all are unlikely to be successful during the selection process and have little or no chance of receiving a job offer
(High Fliers Research 2017)
- As a matter of urgency, more 16-19 year olds must be given opportunities to spend substantial periods in the workplace, undertaking genuine workplace activities, in order to develop the general skills which the labour market demonstrably values...Helping young people to obtain genuine work experience – and, therefore, what the CBI calls 'employability skills' – should be one of the highest priorities for 16-18 education policy in the next few years.
(Professor Alison Wolf, Review of Vocational Education 2011)

You should note that there is no single 'correct' way of planning placements. Much depends on what your school or college can build into the timetable in relation to your needs and the capacity of relevant employers. Your placement could, for example, follow a pattern of once a week for the duration of a term, a block placement of a week or two, or a rotation of shorter placements with a number of employers, so that you can experience different aspects of a sector. You may wish to reflect on the level of work experience that will adequately prepare you for progression to employment, taking into account your abilities, prior attainment, career goals and work readiness.

There are certain important issues to consider in relation to health and safety and insurance. While the employer has the primary responsibility for the health and safety of students on a work experience placement, your school must be satisfied that the employer has assessed the associated risks to workers under 18 on the premises and has suitable and sufficient risk management arrangements in place. For low risk environments, assurance can be gained through a simple conversation with the employer. There is no requirement for a separate risk assessment for work experience students where an employer already employs young workers under 18, as the risks should already have been considered. Where work experience students are the first young workers an employer takes on (or the first for some years), the employer should review the existing risk assessment.

In the rare case where an accident happens on an employer's premises, the employer is normally liable if your school has taken the steps described above to satisfy itself, prior to the placement, that the employer has put in place measures to manage the associated risks in the workplace. As part of the government's Red Tape Challenge, ministers have written to employers confirming that the insurance industry has committed to treat work experience students as employees, so that they will be covered by existing Employers' Liability Compulsory Insurance policies.

If you intend to undertake work experience in the Health Care and Early Years Sector, you must have an enhanced Disclosure and Barring Services (DBS) check before starting on your placement. You will need to factor in the cost of this procedure and also plan for the time it takes to complete an application (which could be up to four weeks).

## Work experience at university

This can take many different forms, ranging from getting a part-time job – possibly with the university itself or the student union – to undertaking a work placement or internship as a stepping stone towards your future career.

Most campuses, for example, offer temporary posts in areas such as administration, promotion and school liaison, bars and hospitality, libraries or computing. Student union posts, on the other hand, are often full-time and would require you to take a year out during or immediately after your degree. You would normally have to campaign to win votes from your fellow students in order to obtain a salaried post as, say, President or as an officer responsible for education, welfare, media or sports and societies.

Any experience of this sort can look impressive on your CV, showing potential employers that you can successfully manage the conflicting demands of academic study and practical work. More specifically relevant to your future career development are issues relating to work placement, work shadowing and internship.

A work placement, rather like school-based work experience, gives you the opportunity to gain hands-on experience of the working world, to learn about teamwork and the importance of meeting deadlines. The difference now, however, is that it should be directly related to the employment sector you wish to enter.

On some courses, a work placement is a compulsory part of the degree; on others you can complete an optional placement as one of your modules. You could still think about arranging a work placement, even if it is not a central feature of your course, and should consult your university careers service for further information.

The length of time you might spend in a work placement will depend on both the employment sector and the individual employer. Some placements are completed during holidays and can last from one to three months; others may involve working one day a week over a period of time.

Certain degree courses require you to spend a full year on a work placement. This is often referred to as a 'sandwich' placement, as it is usually sandwiched between your second and final year at university. Sandwich provision of this type is often found in hospitality and catering degrees. Similar placements commonly occur in engineering, science and construction-related degrees, and may pay a reasonable wage during the course of the year.

You should note that, if the work placement is a compulsory part of your course, it will probably be formally assessed or accredited. Assessment might include the completion of a project, writing a detailed report or using what you have learned on placement as part of another course activity.

Applying for a work placement can be as competitive as applying for a permanent job, and many organisations insist on a formal application detailing why you think you are suitable and what you can bring to the role.

**Work shadowing** lets you observe someone going about their day-to-day job, giving you the opportunity to gain an insight into the nature of a profession and consider whether the role is right for you.

If you want to be a barrister, for example, you can apply for a three- to five-day mini-pupillage to discover first-hand what goes on in court and in chambers. Entry to the profession is so competitive that you may find yourself ineligible for a real pupillage if you have not completed this initial work shadowing experience.

Generally speaking, work shadowing takes place over a few days, although it can sometimes extend into a couple of weeks. It is almost always unpaid and is not usually assessed or accredited as part of a course.

An **internship** can take the form, depending on the sector and the employer, of a four- to ten-week summer vacation placement for undergraduates, or a full year for recent graduates. You gain an insight into how the company works and, if you make a good impression, you stand a great chance of being offered a sought-after graduate job.

You would normally not consider an internship, which gives you direct experience of working in a particular role, unless you are fairly sure that you know what type of job you want to do. Indeed, you will find that many employers nowadays use internships to assess your capabilities and look to recruit employees from their interns rather than simply advertise vacancies. An internship can, therefore, be a significant process in planning your career progression.

Bear in mind that applying for an internship can be as competitive as applying for a permanent job because an internship has become the standard method of recruitment in certain sectors of employment.

There is much debate as to whether you should be paid during an internship. You should be paid at least the UK National Minimum

Wage if you are performing the role of a worker, as part of the commercial operation of a business, with set hours and duties. However, a determining factor may be whether you see the internship as being primarily for your benefit in terms of gaining experience or for the employer in terms of having an extra worker on the premises.

We list below a selection of work experience opportunities aimed primarily at the 16-19 school-based programmes outlined at the beginning of this section. You may find a suitable placement here, although we suggest that you discuss potential opportunities with your work experience coordinator at school and make additional enquiries among local employers.

- In today's competitive graduate recruitment market, work experience schemes have become an integral part of recruiting new graduates. Students applying for paid work placements during their first or second year at university are selected through a very similar recruitment process to that used to recruit graduates. This means that once a placement has been successfully completed, recruiters are able to offer work experience students a graduate position, often a full year before students are due to leave university and several months ahead of employers who only recruit graduates during their final year of study. Many employers also consider that recruiting candidates who have proven their abilities during a work placement to be a more reliable way of employing graduates. (High Fliers UK Graduate Survey 2017)

Another option is to search the online database of the Fair Train Work Experience Quality Standard. Placement providers can have their programmes accredited at Bronze, Silver or Gold Standard, and are committed to offering high quality work experience.

Find out more at: **www.fairtrain.org/work-experience-quality-standard**

## Aviva Work Experience Opportunities 16-18+

Providing some 31 million customers with insurance, savings and investment products, Aviva employs 30,000 people and is one of the leading providers of life and pension plans in Europe. It offers work experience opportunities in cities throughout the UK.

Twitter: @AvivaUKSupport

Web: www.careers.aviva.co.uk

Facebook: www.facebook.com/AvivaUK

## BBC 14-18+

The British Broadcasting Corporation has work experience placements available in just about every area of BBC activity across the UK. Whatever your age and whichever area you're interested in, there could be something right for you, from advertising, charitable work or entertaining to journalism, music or the World Service. All placements are unpaid and can last anything from a few days to four weeks. Competition is fierce, so before you apply you'll need to consider what you can offer and what you'd like to achieve. Are you good with computers? Have you worked in hospital radio or written articles for your local or college magazine? What do you hope to gain from the placement? What are your ambitions for the future? These are the kinds of questions you should be asking yourself.

Twitter: @BBCGetin

Web: www.bbc.co.uk/careers/work-experience

YouTube: www.youtube.com/user/bbccareers

## BDO Summer School — Year 12

BDO LLP, one of the UK's largest accountancy and business advisory firms, offers a two-week programme, during which you will go out on real client assignments, will gain exposure to a variety of industry sectors and will interact with a wide range of senior people, including partners and managers. You must be in Year 12 to be eligible, with Grade B or above in GCSE Maths and English, and predicted to achieve 112 UCAS Points at A level or equivalent.

Twitter: @BDO_Trainees_UK

Web: www.bdoearlyincareer.co.uk/summer-school

Facebook: www.facebook.com/BDOTraineesUK

## Engineering Education Scheme — 16-17

The Engineering Education Scheme gives young people the opportunity to work on a real-life engineering project with a leading organisation across six months. Students work with company mentors to develop employability skills that will help them in their future studies and careers.

Twitter: @TheEDTUK

Web: www.etrust.org.uk/engineering-education-scheme-ees

Facebook: www.facebook.com/TheEDTUK

YouTube: www.youtube.com/user/TheEDTUK

## GlaxoSmithKline 16-18

Leading research-based pharmaceutical and healthcare company GlaxoSmithKline offers an annual work experience programme at its research and development sites in Stevenage, Ware and Harlow during the February half term. It is a week packed with interesting experiences. During your week with them, you learn by doing – on the job – and you can pick up some handy tips from the experienced colleagues around you. Places are available to students in Year 12/13 studying science, IT or engineering. Please note priority is given to students attending schools that are local to the Stevenage, Ware and Harlow areas.

**Twitter:** @GSK

**Web:** http://uk.gsk.com/en-gb/careers

**Facebook:** www.facebook.com/GSK

**YouTube:** www.youtube.com/user/GSKvision

## HSBC 16-18

HSBC Bank runs a structured one- or two-week programme for young people in the later years of full-time secondary, sixth form or college education. You'll experience what it is like to work in retail, commercial, operational or wealth management areas. This will help you decide if you like banking as a career and whether you should join instead of or after going to university.

**Twitter:** @HSBC_UK_Careers

**Web:** www.hsbc.com/careers/students-and-graduates/programmes/uk-work-experience-programme

**YouTube:** www.youtube.com/user/HSBC/custom

## John Lewis Partnership 16-17+

John Lewis offers an opportunity to spend two weeks in a department store, where you can gain an insight into the world of retailing. Placements are normally in selling departments but John Lewis can also accommodate students in other areas of the business if you have a particular interest.

Twitter: @JLPJobs

Web: www.jlpjobs.com

Facebook: www.facebook.com/JLPJobs

## Pinsent Masons 16-18

Pinsent Masons lawyers offer a number of week-long placements across their network of UK offices: Birmingham, Leeds, London and Manchester in England; Aberdeen, Edinburgh and Glasgow in Scotland; or Belfast in Northern Ireland. You will need to be already studying towards either AS or A Levels, or equivalent qualifications. During your week at the firm you will take part in a business exercise, attend information presentations and seminars in addition to shadowing lawyers on real issues. You will spend time in a number of different departments and will be supervised by a trainee solicitor, who will help you gain an appreciation of the skills you need to be a good lawyer.

Twitter: @PMgrads

Web: http://graduate.pinsentmasons.com/programme/school-work-experience

Facebook: www.facebook.com/PMgrads

## PwC Business Insight Week — Year 12

If you are considering a career in business, you can spend a week in a PwC office of your choice during your summer holiday. In addition to developing your employability skills, you will see how PwC advises household brands and global companies on everything from planning for the future to how they can make best use of their technology.

Twitter: @PwC_UK_Careers

Web: www.pwc.co.uk/careers/school-jobs/jobs/work-experience/business-insight-week.html

Facebook: www.facebook.com/PwCCareersUK/

## Year in Industry — 17+

The Year in Industry scheme offers paid, degree-relevant work placements in a year out before or during your university course. With opportunities in all branches of science, technology, engineering, maths and business management, you can undertake real projects and learn how business works. The skills you develop should enhance your university education and maximise your graduate job prospects. Many companies view the scheme as an important part of their recruitment programme, and go on to sponsor placement students through university.

Twitter: @TheEDTUK

Web: www.etrust.org.uk/the-year-in-industry

Facebook: www.facebook.com/TheEDTUK

# Community, Conservation and Heritage Projects in the UK

In this section, you may be able to live at home and work on a local volunteer project, while some schemes insist that being away from home is an essential part of the gap year experience.

## Do-it **All ages**

The Do-it website is a database that currently lists some 1.5 million available volunteering places. The opportunity search page lists placements in your postcode area, for which you can volunteer and apply online. Alternatively, you can use advanced search options to find, for example, causes you are passionate about, or placements that will use your interests and abilities or help you develop new skills. The more information you supply the more likely you are to be matched to organisations looking for willing volunteers.

Twitter: @doituk

Web: https://do-it.org

Facebook: www.facebook.com/DoItUK

## National Citizen Service **15-17**

The National Citizen Service (NCS) is a voluntary personal and social development programme for 15-17 year olds in England and Northern Ireland. It takes place in the spring, summer or autumn coinciding with school holidays. Groups undertake a week-long residential visit, usually to an activity centre for an Outward Bound-style course in the countryside involving physical and team building activities. After this, there is a residential week to gain a taste of independent living and learn a variety of skills for the future. In the third (and sometimes fourth) week, participants plan and deliver a "social action" project in their local community, often to raise awareness of or fundraise for a particular cause. Volunteers have to pay £50 each to take part in the scheme, although there are bursaries for those from low-income households,

Twitter: @ncs

Web: www.ncsyes.co.uk

Facebook: www.facebook.com/ncs

## National Trust — All ages

If you don't want to travel too far but are still looking for a way to make a difference in conserving the environment and the UK's heritage, the National Trust could have something to offer. As a volunteer, you can learn new skills and meet new people while working right at the heart of beautiful buildings, gardens and landscapes. The Trust also runs around 450 Working Holidays every year throughout England, Wales and Northern Ireland, where you could be involved with anything from carrying out a conservation survey to herding goats, painting a lighthouse or planting trees.

Twitter: @nationaltrust

Web: www.nationaltrust.org.uk/volunteer

Facebook: www.facebook.com/nationaltrust

## National Trust for Scotland — 16+

The National Trust for Scotland offers plenty of volunteering opportunities in the heritage sector. It also organises week-long residential working holidays at NT Thistle Camps, where you can get involved in conserving the historic locations under its care. Campers can live and work in some of Scotland's most remarkable and remote places. Whilst Thistle Camps are for over 18s, the trust also runs Trailblazer camps for 16- and 17-year-olds.

Twitter: @N_T_S

Web: www.nts.org.uk/volunteering-jobs

Facebook: www.facebook.com/NationalTrustforScotland

## The Prince's Trust **13-30**

The Prince's Trust helps young people overcome barriers and get their lives working. Through practical support including training, mentoring and financial assistance, the Trust helps 13- to 30-year-olds realise their potential and transform their lives. The main target groups are those who have struggled at school, been in care, been in trouble with the law, or are long-term unemployed.

Twitter: **@PrincesTrust**

Web: **www.princes-trust.org.uk**

Facebook: **www.facebook.com/princestrust**

## Reach **Adult**

Not aimed primarily at school or college leavers, Reach seeks to match the skills of experienced people to the needs of voluntary organisations. Reach recruits and supports people with managerial, technical and professional expertise and places them in part-time, unpaid roles in voluntary organisations that need their help. Volunteers are placed with organisations near where they live, anywhere in the UK.

Twitter: **@ReachSkills**

Web: **https://reachvolunteering.org.uk**

Facebook: **www.facebook.com/ReachSkills**

## vInspired 14-25

vInspired helps you discover the value of volunteering – for yourself and for others. It offers fun, easy-to-access opportunities to get you excited about doing good things. Its online market-place and mobile apps link you with interesting and varied opportunities with almost 2,000 charities across the country. The innovative programmes provide structured support for you to do good and gain new skills in your own way, while earning vInspired Awards to illuminate your CV. Whether you have never volunteered before, or are a veteran volunteer, whether you have an hour or a week to give, vInspired has a project or opportunity for you.

Twitter: @vinspired

Web: https://vinspired.com

Facebook: www.facebook.com/vinspired

YouTube: www.youtube.com/user/vinspiredofficial

## Volunteering Matters 18-35

Volunteering Matters (formerly known as CSV) provides hundreds of full-time volunteering opportunities across the UK that will equip you with life skills and enhance your CV or UCAS application. You will spend 6 to 12 months living away from home, supporting people in need and enabling them to develop or manage their own lives. You can use your skills and develop new ones, test yourself out in new situations, challenge your way of thinking, and make a genuine and positive impact on people's lives. Placements are community-based, supporting a wide variety of people. You may be helping people with physical disabilities or learning difficulties, or supporting elderly people, children or young people.

Twitter: @volunteering_uk

Web: https://volunteeringmatters.org.uk

Facebook: www.facebook.com/volunteeringmatters

YouTube: www.youtube.com/channel/UCoSJ1et8Dwcn_ajk3j5_Z_Q

## International Community and Conservation Projects, Treks and Expeditions

If you want to combine an adventure trek with an environmental, scientific or community project, you may find something suitable here. Expect to pay a substantial participation fee, for which you may have to raise sponsorship. Apart from being one of the best ways to fund your expedition, the experience of having to secure a considerable sum of money will help develop a range of skills before you even leave home! Expedition programmes give you the opportunity to develop other important skills such as communication and time management. You will be building confidence and self-esteem, in addition to learning to lead the expedition team and taking your turn to do so. You should also meet a group of like-minded people and have a lot of fun. Whatever destination you choose, you could end up doing something that will look good on your CV and will be highly valued by universities and employers.

The inspiration for many of the conservation projects listed came from the 'Earth Summit' held in Rio de Janeiro in 1992. This was followed up with the United Nations Conference on Sustainable Development in 2012, also held in Rio and commonly known as Rio+20 or Rio Earth Summit 2012. The initial event was the first global recognition of the urgent need to preserve the diversity of life on earth, culminating in a resolve by 168 countries present to undertake biodiversity surveys of their 'at threat' regions. Many countries committing to this pledge needed assistance with funding and personnel to complete these surveys, leading them to invite some of the organisations listed here to undertake the necessary work. Long-term survival for humans on earth depends on protecting the diversity of organisms, each of which plays a part in maintaining the planet's critical functions.

The aim of community projects is usually to offer humanitarian aid to communities living in poverty, to provide healthcare for the chronically sick, to rebuild homes destroyed by floods and so on. Some schemes ask for volunteers with skills in, say, construction or healthcare.

You should be aware that some projects of this type are derided by critics as mere 'voluntourism', taking jobs away from local people who really need them in order to help wealthy westerners feel good about themselves. There are even suggestions that the problem of fake orphanages in Cambodia, for example, is seriously damaging family life. It is claimed that children are being unnecessarily separated from their families just to take advantage of the resources lavished upon them by well-intentioned but misguided volunteers. According to the pressure group Orphanages No (visit their website at: **www.orphanages.no**), some three quarters of children living in Cambodian orphanages are not orphans at all but simply come from poor families. It is essential that you research providers and projects carefully to be sure that you and your money will bring positive benefit to a community...and, perhaps more importantly, will not be doing any harm.

A reputable organisation should be able to tell you how the fees it charges are being spent, to show you its responsible tourism policy, and to put you in touch both with participants from previous years and with locals on the ground. Look, for example, at reviews on social media, such as Facebook's Better Volunteering community at: **www.facebook.com/BetterVolunteering**

In short, ask the right questions before you make any commitment and plan carefully. Then you can be confident that your skills will be put to good use.

Some agencies listed on the following pages do not specialise in any one particular type of project. Activities may be paid or unpaid, while some are predominantly study-based. Typical offerings might include Volunteering (Community Development, Construction, Conservation, Teaching), Internships (Media, Marketing, Health, Tourism, Sports Coaching), Tours (Cultural, Ecological, Humanitarian, Railway) and Jobs (all over the world). We have also listed some associations – such as the Year Out Group – which act as an umbrella group for several separate providers.

## Africa and Asia Venture (AV) 18-25

Want to combine four to five months of travel, safaris, friendship and fun in a year out with teaching or coaching sports, working with local communities or conservation work? You would start with a four-day in-country training course, then spend three to four months at your project, followed by one month of travel and safari. The latter could include seeing wildlife in African National Parks, visiting temples and palaces in Nepal, or adventure in Mexico. In Thailand, you would teach among the hill tribes for two months, followed by one month's project work, three weeks of travel/adventure opportunities and a five-day open water diving course.

Countries: China, India, Kenya, Malawi, Nepal, South Africa, Tanzania, Thailand, Uganda

Twitter: @AVentureUK

Web: www.aventure.co.uk

Facebook: www.facebook.com/AVentureUK

YouTube: www.youtube.com/user/AfricaAsiaVenture/videos

## African Conservation Experience 17+

African Conservation Experience claims to be one of the most experienced organisations for conservation placements in Southern Africa. It can offer you the chance to work on game and nature reserves alongside conservationists, zoologists, wildlife vets and reserve managers. The organisation welcomes volunteers from all backgrounds, with no previous experience necessary, from the age of 17 upwards. Volunteer Placements are from two weeks to three months, and you can combine two or more projects in one trip. You could join a placement as part of a gap year, in a summer break from school or university, or as part of a career break or sabbatical.

Countries: Botswana, Namibia, South Africa, Zimbabwe

Twitter: @AfricanConsExp

Web: www.conservationafrica.net

Facebook: www.facebook.com/AfricanConservationExperience

## Agape Volunteers 17+

Agape Volunteers provide opportunities in a wide range of well-established, affordable volunteer placements to support humanitarian work in Kenya, Ghana, South Africa and Tanzania. They organise opportunities to teach in a school, work in an orphanage, develop sports education programmes and work in conservation. Medical students can join one of the hands-on placements in an African hospital or clinic. Placements are geared towards making a difference and volunteers are guaranteed hard work and a life-changing trip.

**Countries:** Ghana, Kenya, South Africa, Tanzania

**Twitter:** @AgapeVolunteer

**Web:** www.agape-volunteers.com

**Facebook:** www.facebook.com/AgapeVolunteers

## Bamboo-GVN 18+

The Bamboo Project powered by GVN (Global Volunteer Network) is a newly-formed partnership offering opportunities in independent volunteering and adventure travel. It currently provides volunteer projects focusing on childcare, community development, conservation, construction, education and medical needs across South America, Asia and Africa.

**Countries:** Cambodia, Costa Rica, Everest, Guatemala, Hawaii, India, Nepal, Peru, Tahiti, Thailand, Uganda, Vietnam

**Twitter:** @WeAreBamboo

**Web:** www.bamboo-gvn.org

**Facebook:** www.facebook.com/WeAreBamboo

## Big Beyond 18+

Big Beyond organises ethical volunteer placements in Africa for adventurous individuals looking for experiences far beyond the norm, off the beaten path and with the chance to become immersed in another culture. Volunteer placements are always tailored to your skills, interests and level of experience, as well as the needs of the local community. There are currently project sites in Uganda and Ethiopia. Volunteer involvement ranges from education, enterprise and environment to health- or culture-related projects.

Countries: Kenya, Tanzania, Uganda, Zanzibar

Twitter: @BigBeyond

Web: www.bigbeyond.org

## Blue Ventures 17+

A not-for-profit organisation dedicated to facilitating projects and expeditions that enhance global marine conservation and research, Blue Ventures coordinates expeditions consisting of scientists and volunteers, working hand-in-hand with local biologists, governmental departments and communities, to carry out research, environmental awareness and conservation programmes at threatened marine habitats around the world.

Countries: Belize, Madagascar, Timor-Leste

Twitter: @BlueVentures

Web: https://blueventures.org

Facebook: www.facebook.com/blueventures

## British Exploring 16+

British Exploring organises extreme adventure and conservation expeditions in remote, wild environments. You could find yourself monitoring climate change in the Arctic, measuring biodiversity in the jungle or investigating human impact on the environment in mountainous regions. The aim is always to develop the confidence, teamwork, leadership and spirit of adventure and exploration of all expedition members.

**Twitter:** @Brit_exploring

**Web:** www.britishexploring.org

**Facebook:** www.facebook.com/BritishExploringSociety

## BUNAC 18+

BUNAC offers a wide range of working holidays in Australia, New Zealand, Canada and the USA, including the Summer Camp Exchange USA programme, and internships and volunteering programmes on marine, conservation, teaching and wildlife projects throughout the world from Fiji to South Africa. These are open to young people in the UK, the USA and Ireland. Programmes last from 12 days up to two years.

**Countries:** Australia, Cambodia, Canada, China, Ecuador, India, Japan, Mexico, Nepal, Philippines, New Zealand, South Africa, Thailand, USA, Vietnam

**Twitter:** @BUNAC_UK

**Web:** www.bunac.org/uk

**Facebook:** www.facebook.com/BUNACworkabroad

## Camps International 18-25

Camps International operates a permanent network of international volunteer camps in the heart of rural communities and wildlife conservation areas in Kenya, Tanzania, Borneo, Cambodia, Peru and Ecuador. Camps offer you the chance to make a difference by volunteering in over 100 community, wildlife and conservation projects, bringing revenue and assistance to rural communities that would otherwise not benefit from tourism.

Countries: **Borneo, Cambodia, Costa Rica, Ecuador, Kenya, Malawi, Nicaragua, Peru, Tanzania, Uganda**

Twitter: **@CampsInt**

Web: **http://campsinternational.com**

Facebook: **www.facebook.com/campsinternational**

## Changing Worlds 17+

Changing Worlds is a travel organisation that provides travel experiences for individuals looking to take part in placements overseas during a gap year, career break or as part of a constructive trip abroad. Placements range from career development, internships and work experience, courses and qualifications, paid work options and community volunteering to cultural and adventure tours.

Countries: **Argentina, Australia, Cambodia, Canada, China, Costa Rica, Ecuador, Fiji, Germany, Ghana, India, Indonesia Kenya, Malaysia, New Zealand, Philippines, St Lucia, South Africa, Sri Lanka, Thailand, Vietnam**

Twitter: **@Changing_Worlds**

Web: **www.changingworlds.co.uk**

Facebook: **www.facebook.com/changingworlds.gapyears**

## Concordia — 16-30

In addition to UK farm placements, Concordia offers an extensive international volunteer programme. Short-term projects bring together individuals from around the world to participate in two- to four-week projects in Western, Eastern and Central Europe, North America, Latin America, North Africa, Japan and South Korea. Medium-term projects usually last between one and six months, although they occasionally last for a whole year. Every spring Concordia posts its new summer programme online for the forthcoming year.

Countries: Worldwide

Twitter: @ConcordiaVol

Web: www.concordiavolunteers.org.uk

Facebook: www.facebook.com/ConcordiaVolunteers

## Coral Cay Conservation — 18+ (16+ if a school group)

Coral Cay Conservation (CCC) is a not-for-profit conservation organisation, sending teams of volunteers to survey some of the world's most endangered coral reefs and tropical forests. Its mission is to protect these crucial environments by working closely with the local communities who depend on them for food and livelihood. CCC currently has a marine expedition in the Philippines, and a terrestrial and marine expedition in Montserrat. The organisation is financed through a combination of grants, and volunteers who pay to participate in an expedition for anything from one week upwards. Volunteers do not need to have any science or dive training as all of the necessary training is covered on site.

Countries: Philippines

Twitter: @CoralCay

Web: www.coralcay.org/volunteer

Facebook: www.facebook.com/CoralCayConservation

## Dragoman Overland 18+

Dragoman has been running adventure trips for over 30 years across the four continents of Africa, Asia, North and South America. Its overland tours range from 9 to 235 days in length. The shorter two- or three-week trips offer a snapshot of a country or region, giving travellers a short, sharp adventure travel experience. A simple break from everyday life can get you off the tourist trail and take you on a perspective-changing journey. On one of the longer tours you could cover a whole continent. The iconic Istanbul to Beijing overland follows in the footsteps of the ancient Silk Road caravans, or you can join Dragoman on a trans-Africa expedition from Cape Town to Cairo.

Twitter: @DragomanTravel

Web: www.dragoman.com

Facebook: www.facebook.com/dragomantravels

## Earthwatch 15-18

Earthwatch Teen Expeditions are designed specifically and exclusively for 15- to 18-year-olds. Working with projects all around the world, they're hands-on, engaging and meaningful – providing unrivalled opportunities to undertake vital, peer-reviewed scientific field research under the supervision of skilled research teams in a professional setting. Whether you're interested in helping conserve the Amazon Basin's pink river dolphins, snorkel and survey reefs in Belize, or dig for cultural treasures in a Roman villa, Earthwatch has an expedition for you. If you are 18, you have the option of joining a Teen Expedition or a regular expedition as an adult.

Twitter: @earthwatch_Eur

Web: https://earthwatch.org.uk

Facebook: www.facebook.com/Earthwatch

## Ecoteer Responsible Travel 18+

Ecoteer establishes project sites working alongside local communities to create mutually beneficial outcomes. At each project site, Ecoteer has offered education and alternative sources of income for the local communities, with up to 35% of volunteer fees being given to them for services rendered. The Ecoteer programmes also provide donations and manpower to the organisations that they work with, such as the Malaysian Conservation Alliance for Tigers (MYCAT) and Save the Caves of Merapoh.

Countries: Indonesia, Malaysia

Twitter: @ecoteer

Web: https://ecoteer.com

Facebook: www.facebook.com/ecoteer

## Edge of Africa 18+

Edge of Africa is an award-winning volunteer project coordinator for gap year placements, work experience and volunteer internships in Africa. Its conservation work includes lemur research, elephant and cheetah care and black rhino breeding, while there are also volunteer projects in the community and internships in such fields as criminology and marketing in South Africa.

Countries: South Africa

Twitter: @edgeofafrica

Web: www.edgeofafrica.com

Facebook: www.facebook.com/EDGEofAFRICA

## Experiment in International Living (EIL) 16+

EIL provides programmes for students who want to connect deeply and engage meaningfully with the richness and complexities of another country. Participants explore the host country through hands-on experience in local communities and through the lens of a specific theme. The programmes are designed to equip you not only with essential cultural and, in many cases, language skills, but also with a deeper awareness of, and sensitivity to, critical global issues shaping the diverse communities and regions visited.

Countries: Worldwide

Twitter: @TheExperimentWL

Web: www.experiment.org

Facebook: www.facebook.com/experimentabroad

## Frontier 17+

Frontier is a non-profit conservation and development organisation dedicated to safeguarding biodiversity and ecosystem integrity and building sustainable livelihoods for marginalised communities in the world's poorest countries. It hosts a myriad of global conservation projects, offering 'hands-on' fieldwork, which benefits endangered tropical wildlife and their ecosystems and directly assists developing countries rich in biodiversity but poor in the capacity to manage natural resources.

Countries: Worldwide

Twitter: @FrontierGap

Web: www.frontier.ac.uk

Facebook: www.facebook.com/frontierofficial

## G Adventures 18+

G Adventures arranges a variety of adventure holidays around the world, giving you the opportunity to step off the beaten track and experience authentic accommodation and local transportation to bring you face-to-face with the world's most fascinating cultures, customs and awe-inspiring wildlife. Adventures include expeditions to the Arctic and Antarctica that call out to a select breed of traveller. Are you one of them? Set sail aboard the G Expedition and explore these harsh and unbelievably rewarding frozen frontiers. Visit North Africa and the Middle East, where holy relics, ancient ruins and other reminders of yesterday may start to explain why travellers come to this part of the world.

Countries: **Worldwide**

Twitter: **@gAdventures**

Web: **www.gadventures.com**

Facebook: **www.facebook.com/gadventures**

## Gapforce 17+

Gapforce offers overseas structured gap year programmes, volunteer projects and extreme expeditions. Gap year programmes can vary from marine conservation, adventure travel, community volunteering, wildlife conservation, outdoor training and multi-country experiences. If you want to help a worthy cause there are summer volunteer programmes and for intrepid explorers there is expedition leader training.

Countries: **Africa, Asia, Europe, South America, Central America and the Caribbean, Australia and the South Pacific**

Twitter: **@gapforce**

Web: **https://gapforce.org/gb**

Facebook: **www.facebook.com/GapforceUK**

## GapGuru 18+

GapGuru is a gap year specialist offering a wide range of volunteer, travel and internship opportunities across Asia, Africa, South America and Europe. With GapGuru you could be working as a medical intern in India, building schools in Tanzania or even trekking up Mount Kilimanjaro! GapGuru gives you the opportunity to discover new countries and cultures, to meet new people and immerse yourself in local communities. Volunteers work in underprivileged communities, helping those in genuine need, while building their own skills, confidence and experience.

Countries: **Cambodia, India, Nepal, Peru, Romania, Tanzania, Thailand.**

Twitter: **@GapYearGuru**

Web: **www.gapguru.com**

Facebook: **www.facebook.com/GapGuru**

## Gapwork 17+

The Gapwork website aims to provide everything you might need to know about taking a gap year and working abroad. It offers gap year and backpacking information, including news of gap year jobs, travel destinations and volunteering ideas, as well as a comprehensive planning section. Whether it's skiing in the Rocky Mountains, volunteering in a South African safari park or tracking river dolphins in the Amazon, there could be something suitable in the Gap Year Ideas section.

Countries: **Worldwide**

Twitter: **@GapworkJobs**

Web: **www.gapwork.com**

Facebook: **www.facebook.com/gapwork**

## Gap Year South Africa 18+

Gap Year South Africa works extremely closely with under-resourced local South African communities. South African Gap Year projects contribute to social and economic development and address priority areas such as education, health, and social cohesion. Gap Year South Africa offers 3-week, 5-week, 3-month and 5-month project options in Cape Town, and volunteer projects include Teaching and Education, Sports Coaching, Performing Arts, HIV/AIDS Awareness, Care Work, Medical and Veterinary projects and Environmental Awareness, Scuba Diving, Marine Conservation and Surfing Projects.

Countries: South Africa

Twitter: @GapYear_SA

Web: http://gapyearsouthafrica.com

Facebook: www.facebook.com/gapyearsouthafrica

## Global Choices 17+

Global Choices offers a wide variety of programmes for students, working with partner organisations all over the world to arrange job placements and volunteering opportunities. You could spend some time working in a hotel in Argentina, teaching English in Brazil or Spain, or you could travel to the US and earn a modest wage. You could try a working holiday programme in Australia or a summer in Camp USA. How about helping a disadvantaged community or taking part in a conservation programme? Global Choices has a vast range of programmes all over the world to help individuals enhance their professional and personal skills.

Countries: Worldwide

Twitter: @GlobalChoices

Web: www.globalchoices.co.uk

Facebook: www.facebook.com/GlobalChoices

## Global Volunteer Projects 17+

This organisation offers a variety of placements and projects across the world including medical placements, teaching projects, HIV awareness, journalism placements, conservation projects and orphanage placements. You could find yourself working at a Ghanaian TV station, at an orphanage in Cambodia or helping to preserve endangered sea turtles on the Pacific coast of Mexico.

**Countries:** Cambodia, China, Ghana, India, Mexico, Tanzania

**Twitter:** @GVPteam

**Web:** www.globalvolunteerprojects.org

**Facebook:** www.facebook.com/GlobalVolunteerProjects

## GVI Foundations 15-17, 18+

Global Vision International (GVI) offers more than 150 different volunteer abroad projects in 25 countries all over the world. By taking part, you can play a crucial role in a wide range of conservation, education and community projects in Africa, South America, Asia, Europe and Central America. GVI has created specially designed projects for young adults, a blend of volunteering, cultural and adventure activities.

**Countries:** Worldwide

**Twitter:** @GVIWORLD

**Web:** www.gvi.co.uk/volunteer-abroad

**Facebook:** www.facebook.com/GVIfans

## Habitat for Humanity 16+

An international aid charity working to end poverty housing around the world, Habitat for Humanity works in over 3,000 communities in 70 countries. Volunteer teams work alongside local communities and housing beneficiaries to provide them with a safe and decent place to live. Projects include housing orphans in Lesotho, funding small business loans in Tajikistan, setting up refugee camps in Macedonia and creating clean water supplies in Delhi.

Countries: **70+ countries around the world**

Twitter: **@HabitatFHGB**

Web: **www.habitatforhumanity.org.uk**

Facebook: **www.facebook.com/habitatforhumanitygb**

## i-to-i /Love TEFL 17+

The recent merger of two specialist organisations in Teaching English as a Foreign Language (TEFL) has created opportunities for participants to obtain a TEFL qualification before embarking on a job, an immersive internship programme or an adventure travel project. Initial TEFL training requires a minimum of 120 hours, all of which can be carried out online, or you can add 20 hours' classroom learning to gain practical experience.

Countries: Worldwide

Twitter: @i_to_iTEFL

Web: www.i-to-i.com

Facebook: www.facebook.com/itoitefl

## International Voluntary Service (IVS GB) 18+

A peace organisation working for the sustainable development of local and global communities throughout the world, IVS GB is the British branch of Service Civil International, a worldwide network of like-minded voluntary organisations promoting peace and justice through voluntary work. By taking part in an international voluntary project, you will be working and living alongside other volunteers from all over the world and contributing to local community development. There are hundreds of projects to choose from, including environmental conservation on beaches in Morocco, help at a centre for children with disabilities in Latvia, work with elderly people in mountain villages in Japan, a community theatre in the Czech Republic, or youth work in Russia.

Countries: Worldwide

Twitter: @IVSGB

Web: www.ivsgb.org

Facebook: www.facebook.com/ivsgb

## Lattitude Global Volunteering 17-25

Lattitude organises voluntary work overseas, and similar exchange voluntary opportunities for overseas nationals in the UK. Lattitude volunteers work alongside staff in such roles as foreign language assistants, helping with general activities in schools, caring for the disadvantaged, or in outdoor education and conservation work. Whatever the nature of your placement, it will always be a challenge. You could be working in an environment different from anything you've ever experienced, so you need to adapt to your responsibilities with maturity. As well as making a real difference to the lives of others, you'll certainly learn a lot about yourself.

Countries: Argentina, Australia, Canada, China, Ecuador, Fiji, Ghana, Ireland, Japan, Malawi, New Zealand, Poland, Spain, Vietnam, UK

Twitter: @LattitudeUK

Web: www.lattitude.org.uk

Facebook: www.facebook.com/globalvolunteering

## Love Volunteers 15-17, 18+

Love Volunteers offers 'not-for-profit programme fees' for fun, rewarding, safe and affordable volunteering opportunities in developing countries around the world. Love Volunteers has a variety of programmes available, whether you're planning a week helping a local community as you travel around a country, want to spend six months learning the local language or are putting your gap year to good use.

**Countries:** Africa, Asia, Australasia, Central America, Europe, Middle East, South America, The Caribbean

**Twitter:** @LoveVolunteers

**Web:** www.lovevolunteers.orgFacebook:

**Facebook:** www.facebook.com/LoveVolunteers

## Madventurer 17+

Madventurer rural projects focus on building basic infrastructure to assist local community development. The key purpose is youth development and the provision and improvement of health and education through work on schools, clinics, toilets, water storage tanks, community centres and sanitation facilities. Venturers also have the opportunity to teach English and other subjects in local primary schools, as well as getting involved in extracurricular activities such as sports, art and drama. As a volunteer, you become part of the Mad Tribe. Each year there is a Mad World Ball in Newcastle upon Tyne, for reunions and reminiscing about the time you've weathered and treasured together.

**Countries:** Fiji, Ghana, Kenya, Peru, South Africa, Sri Lanka, Tanzania, Thailand, Togo, Uganda

**Twitter:** @Madventurer

**Web:** www.madventurer.com

**Facebook:** www.facebook.com/Madventurer

## Oasis Overland 17+

Oasis Overland is an adventure travel company with a huge number of overland trips in several continents, including Africa, the Middle East, Latin America and Asia. Oasis Overland trips can take up to 24 people, mostly in their purpose-built vehicles, and vary in length up to 38 weeks. Your expedition could involve anything from 9 days in Egypt to 38 weeks across Africa.

Countries: Worldwide

Twitter: @OasisOverland

Web: www.oasisoverland.co.uk

Facebook: www.facebook.com/OasisOverland

## Outreach International 16+

This specialist organisation has a wide variety of projects, all of them small, grassroots initiatives working with communities where volunteer work can make a big difference. Placements last between one month and twelve months and could involve anything from working with children – in schools or orphanages – to helping out at an animal rescue centre. Other projects include rainforest and marine conservation, teaching English or art and craft, or medical-related projects, such as community medicine and physiotherapy.

Countries: Cambodia, Costa Rica, Ecuador, Galápagos Islands, India, Kenya, Mexico, Nepal, Sri Lanka, Tanzania

Twitter: @Outreachint

Web: www.outreachinternational.co.uk

Facebook: www.facebook.com/outreachinternational.uk

## Overseas Job Centre — All ages

This site is a guide to living and working abroad, seasonal jobs, working holidays, careers breaks, gap years, volunteer work and long-term world travel. It lists volunteer work opportunities around the world, from tourism and catering to teaching or working with animals. It gives advice on planning your gap year as well as a useful list of websites for reference and further information.

Countries: **Worldwide**

Twitter: **@WorkingTravellr**

Web: **www.overseasjobcentre.co.uk**

Facebook: **www.facebook.com/workingtraveller**

## Oyster Worldwide — 18+

Oyster is an experienced gap year and travel organisation that has been sending volunteers and paid workers to diverse destinations since 1999. Volunteering projects include: childcare, teaching English, sports coaching, animal welfare and environmental conservation. Paid work opportunities include: ski instructing, hospitality work, farm work and teaching English. Oyster also offers medical, veterinary and business internship experiences.

Countries: **Australia, Borneo, Brazil, Canada, Chile, China, Costa Rica, Ecuador, France, Galápagos Islands, India, Laos, Malaysia, Namibia, Nepal, New Zealand, Romania, South Africa, Sri Lanka, Tanzania, Thailand, Uganda**

Twitter: **@OysterWorldwide**

Web: **www.oysterworldwide.com**

Facebook: **www.facebook.com/Oyster.Gap.Year**

## Pacific Discovery 18-24

Pacific Discovery offers inspiring gap year, experiential and educational travel programmes to the most amazing places on earth. Its award-winning programmes blend meaningful and challenging travel, cultural immersion, personal and leadership development, outdoors and wilderness exploration, volunteer and community service projects, ethical travel and sustainability focus.

**Countries:** **Australia, Cambodia, Cuba and Central America, Ecuador, Fiji, Galápagos Islands, Laos, Nepal, New Zealand, Peru, Thailand, Tibet, Vietnam**

Twitter: @pacificdiscovry

Web: www.pacificdiscovery.org

Facebook: www.facebook.com/PacificDiscovery

## People and Places: responsible volunteering 18+

A winner at the 2013 World Responsible Tourism Awards, the ethos of People and Places is that 'There are still way too many unethical or poorly managed volunteer organisations operating. We believe volunteering can be a win-win for volunteers and the community they seek to serve – but people must be put in the right places. If you can reserve a place on a volunteer programme simply by providing contact details and your credit card number, and if you don't know where exactly you'll be working and what you'll be doing, ask yourself: should I be volunteering with that organisation?'

Countries: Cambodia, Gambia, India, Kenya, Madagascar, Morocco, Nepal, Peru, St Lucia, South Africa, Swaziland

Twitter: @pandpvolunteer

Web: http://travel-peopleandplaces.co.uk

Facebook: www.facebook.com/Responsible-volunteering-people-and-places 109179132471692

## PoD Volunteer 16+

PoD is a non-profit organisation arranging ethical and supported volunteering opportunities around the world. It works with long-term projects where there is genuine benefit to local communities. Whether it is for one week or six months, volunteer jobs can vary from assisting with animal care and rescue, helping at children's homes, teaching English, building, filming and photography, supporting community development or joining an active research team to assist with conservation work.

Countries: **Belize, Cambodia, Costa Rica, Ghana, India, Madagascar, Namibia, Nepal, Peru, South Africa, Thailand,**

Twitter: **@podvolunteer**

Web: **www.podvolunteer.org**

Facebook: **www.facebook.com/podvolunteer**

## Project Trust 17-19

Based on the Isle of Coll, Project Trust specialises in 8- or 12-month volunteer placements, giving you plenty of time to explore your new surroundings whilst working in the local community. You can choose from over 20 different countries, spending a year living, working and travelling in Africa, Asia, Latin America or the Caribbean with a wide variety of work and a diverse range of cultures.

Countries: **Botswana, Cambodia, Chile, China, Dominican Republic, Ghana, Guyana, Honduras, Hong Kong, India, Japan, Malawi, Malaysia, Nepal, Peru, Senegal, South Africa, Swaziland, Thailand, Zambia**

Twitter: **@ProjectTrustUK**

Web: **www.projecttrust.org.uk**

Facebook: **www.facebook.com/ProjectTrustVolunteers**

## Projects Abroad 16+

With a wide range of projects, including teaching, care, conservation, medicine and healthcare and journalism, Projects Abroad organises overseas voluntary work placements designed specifically for the communities where it works. The teaching projects focus on conversational English teaching and don't require TEFL qualifications. You could teach in Africa, Asia, Latin America or Eastern Europe, as part of a project in a school, university or orphanage. In journalism, you could work on a Chinese, Indian, Ghanaian, Mexican, Moldovan, Mongolian, Romanian or Sri Lankan newspaper or work at a radio station in Ghana, Senegal or Mexico, or even a TV station in Mongolia. A range of medical, conservation, animal care, business and sports internships are also available.

Countries: **Worldwide**

Twitter: **@Proj_Abroad_UK**

Web: **www.projects-abroad.co.uk**

Facebook: **www.facebook.com/projectsabroaduk**

## Quest Overseas 17+

Specialists in Africa and South America, Quest offer 'Combined Gap Expeditions', in which you can learn a language, work on a community or conservation project and then explore the best of the surrounding countries. This could lead, for example, to Community Development work in Tanzania or a Game Reserve project in Swaziland, together with exploration of southern Africa, or an Animal Sanctuary project in Bolivia, together with exploration of the Andes.

Countries: **Bolivia, Peru, Rwanda, Malawi**

Twitter: **@QuestOverseas**

Web: **www.questoverseas.com**

Facebook: **www.facebook.com/questoverseas**

## Rail Adventure across Europe — All ages

InterRail offers European train passes and information on how to travel around Europe by train, including planning your journey, timetables, maps and passes. The most popular product is the InterRail Global Pass which gives you flexible train travel in 30 countries. InterRail offer a special youth discount up to the age of 25.

Countries: **Austria, Belgium, Bosnia-Herzegovina, Bulgaria, Croatia, Czech Republic, Denmark, FYR Macedonia, Finland, France, Germany, Great Britain, Greece, Hungary, Ireland, Italy, Luxembourg, Montenegro, Netherlands, Norway, Poland, Portugal, Romania, Serbia, Slovakia, Slovenia, Spain, Sweden, Switzerland, Turkey.**

Twitter: **@InterRail**

Web: **www.interrail.eu**

Facebook: **www.facebook.com/interrail**

## Raleigh International — 17-24

The Raleigh overseas programme enables participants from all over the world and from all backgrounds to undertake a blend of mental and physical challenges. The full 10-week programme consists of three distinct project phases – sustainable community and environmental projects plus an adventure phase. There are also five-week programmes, which combine your choice of either a community or environmental project with a team-based adventure challenge.

Countries: **Borneo, Costa Rica, Nepal, Nicaragua, Tanzania**

Twitter: **@Raleigh_**

Web: **https://raleighinternational.org**

Facebook: **www.facebook.com/RaleighInternational**

## Real Gap Experience 17+

Real Gap Experience offers a comprehensive range of volunteering, paid work, sports, adventure travel, language courses and career breaks in over 40 countries, from two weeks to two years. The company can organise a complete gap year away unique to you (including flights and insurance) with a strong emphasis on security and safety. All Real Gap Experience advisers have taken their own gap year and will provide help and support to plan yours.

Countries: Worldwide

Twitter: @RealGapExp

Web: www.realgap.co.uk

Facebook: www.facebook.com/realgapexperience

## Restless Development 18-28

If you are passionate about changing lives and want to make a difference to the community that you work with, you can volunteer to work for five to eleven months in communities around the world. You might find yourself helping vulnerable young people protect themselves against HIV, open a library or a youth centre, construct a smokeless stove or establish a recycling programme.

Countries: India, Nepal, Sierra Leone, South Africa, Tanzania, Uganda, UK, USA, Zambia, Zimbabwe

Twitter: @RestlessDev

Web: http://restlessdevelopment.org

Facebook: www.facebook.com/restlessdev

## Royal Geographical Society 16+

Every year the Society supports between 40 and 50 teams of students and researchers to get into the field with a Geographical Fieldwork Grant, the Society's longest-running grant scheme. The three independent travel grants support challenging and inspiring geographical journeys and expeditions and are worth up to £3,000. Full details of the different grants available can be found on the website. Fully funded places are available on a five-day fieldwork summer school for AS geography students at the end of Year 12. The courses give students who would not normally have such opportunities a chance to experience high quality residential fieldwork.

**Twitter:** @RGS_IBG

**Web:** www.rgs.org/in-the-field

**Facebook:** www.facebook.com/RGS.IBG

## SEED Madagascar 18+

SEED (Sustainable Environment, Education and Development) Madagascar is the new name of Azafady, a specialist organisation working for a better future for people, communities, and the environment in the town of Fort Dauphin and the Anosy region of south-east Madagascar. The approach is one of partnership with local communities to alleviate poverty and support environmental conservation. Integrated sustainable development initiatives support some of the world's most vulnerable people in threatened, irreplaceable environments.

**Countries:** Madagascar

**Twitter:** @SEEDMadagascar

**Web:** http://madagascar.co.uk

**Facebook:** www.facebook.com/SEEDMadagascar

## The Leap 17-27

The Leap offers adventurous volunteering team placements in Africa, South America and Asia, along with internship opportunities in Tanzania and Ecuador. Its projects aim to broaden your mind and your muscles and surpass your expectations.

Countries: **Bahamas, Borneo, Cambodia, Costa Rica, Cuba, Ecuador, Kenya, Madagascar, Namibia, Nepal, Peru, South Africa, Tanzania, Venezuela**

Twitter: **@GapYearExperts**

Web: **http://theleap.co.uk**

Facebook: **www.facebook.com/GapYearExperts**

## Tourdust 18+

Tourdust is a specialist adventure tour operator for independent travellers looking to book activity holidays, adventure holidays and tours from expert local guides all around the world. Experiences offered by Tourdust include Inca Trail treks and trekking in Morocco's Atlas Mountains, alongside a host of other trips such as safaris in the Masai Mara and sea kayaking in the Aegean.

Countries: **Botswana, Croatia, Ethiopia, Kenya, Morocco, Namibia, Peru, South Africa, Sweden, Tanzania, Turkey, USA, Zambia**

Twitter: **@Tourdust**

Web: **www.tourdust.com**

Facebook: **www.facebook.com/tourdust**

## Travellers Worldwide 17+

Play a meaningful role in helping to preserve threatened and endangered species. You don't need any qualifications, just a love of animals. You can work with monkeys or orang-utans, elephants, pandas, turtles, horses, plants, dolphins, whales and the Great White Shark, leopards, rhino, tortoise, coral, reptiles, birds or fish. The type of work varies and can include: interaction and play, field research, safaris, washing, feeding, observation, releases into the wild, diving, trekking, darting, educating, photographing, building, painting and more depending on the requirements of the project.

Countries: **Argentina, Australia, Bali, Brazil, Cambodia, China, Costa Rica, Ecuador, Ghana, India, Indonesia, Madagascar, Malaysia, New Zealand, Peru, South Africa, Sri Lanka, Thailand, USA, Zambia, Zimbabwe**

Twitter: **@TravellersWW**

Web: **www.travellersworldwide.com**

Facebook: **www.facebook.com/travellersworldwide.projects**

## TrekAmerica 18+

TrekAmerica's distinctive style of adventure travel offers something different to your ordinary guided tour. It organises small group adventures across the USA, Canada, Alaska and Central America. Every year thousands of young people from all over the world find that its small group tours offer an easy and dependable way to explore this vast area.

Countries: **Belize, Canada, Costa Rica, Guatemala, Honduras, Mexico, Nicaragua, Peru, USA**

Twitter: **@trekamerica**

Web: **www.trekamerica.co.uk**

Facebook: **www.facebook.com/trekamerica**

## Twin Work and Volunteer — All ages

Twin Work and Volunteer provides a wide range of programmes allowing people of all backgrounds and ages the opportunity to participate in community and conservation volunteering projects around the world, together with internships and paid work placements. In a gap year abroad you could save endangered animals, build schools or work in an orphanage. You may want to volunteer for a community or conservation project or live and work abroad on a paid working holiday. Or you could add valuable work and life experience to your CV with an internship abroad or in the UK from four weeks to a year.

Countries: **Worldwide**

Twitter: **@Work_Volunteer**

Web: **www.workandvolunteer.com**

Facebook: **www.facebook.com/TwinWorkandVolunteer**

## VentureCo — 17-20

VentureCo's multi-phase travel programmes feature journeys of discovery on horseback, although there are also activities for non-riders. The ventures are a combination of complementary phases. For example, in South America you would combine a Spanish language phase with a Project phase and an Expedition phase to make one venture. Each phase reveals a different aspect of your host country and together they produce one memorable travel experience.

Countries: **Argentina, Bolivia, Botswana, Bulgaria, Chile, Columbia, Ecuador, Guyana, Iceland, India, Jordan, Lesotho, Mongolia, Morocco, Namibia, Patagonia, Peru, Romania, Spain, Swaziland, Uganda, USA**

Twitter: **@VentureCoUK**

Web: **www.ventureco-worldwide.com**

Facebook: **www.facebook.com/Ventureco**

## Voluntary Service Overseas (VSO ICS) 18-25

The VSO International Citizen Service brings together people from all backgrounds to fight poverty overseas. Funded by the UK government, it offers the opportunity to contribute to development projects by working for three months in Africa, Asia or Latin America, alongside young in-country volunteers, whilst gaining transferable skills and experience. You don't need cash, skills or qualifications to volunteer with VSO ICS – just the ambition to make a difference.

Countries: **Bangladesh, Cambodia, Ghana, Ethiopia, India, Kenya, Lesotho, Malawi, Mozambique, Myanmar, Nepal, Nigeria, Pakistan, Papua New Guinea, Philippines, Rwanda, Sierra Leone, South Africa, Swaziland, Tanzania, Uganda, Zambia, Zimbabwe**

Twitter: @VSO_Intl

Web: www.vsointernational.org

Facebook: www.facebook.com/vso.international

## Volunteer Action for Peace 18+

Volunteer Action for Peace (VAP) is a UK-based charity organisation which works towards creating and preserving international peace, justice and human solidarity for people and their communities. Through a range of working projects both in the United Kingdom and around the world, VAP provides volunteers with opportunities to work together with people from around the globe and in partnership with local groups to enhance and empower communities. Opportunities vary from two to four weeks to six-plus months.

Countries: Worldwide

Twitter: @action_peace

Web: www.vap.org.uk

Facebook: www.facebook.com/pages/Volunteer-Action-for-Peace/193838147321030

## Volunteering India 18+

Volunteering India provides safe, affordable and meaningful volunteer programmes in India. It offers a variety of volunteer work, cultural exchange, internships, and gap years, where volunteers can choose to work in New Delhi, Palampur/Dharamsala and South India. You could work with orphans, in female empowerment, Health/HIV, teaching English, summer volunteer work or a street children volunteer programme.

Countries: India

Twitter: @indiavolunteer

Web: www.volunteeringindia.com

Facebook: www.facebook.com/VolunteeringIndia

## Wind, Sand and Stars 16-23

Wind, Sand and Stars are desert specialists. Working closely with local tribal communities they offer authentic, personal and close-up experiences. The trips have combined desert adventures, mountain treks, camel safaris, field trips, school journeys, silent retreats, pilgrimage to historical religious sites, charity projects and more. Whether travelling through the deserts of Arabia or highlands of Ethiopia, they have built strong working relationships with the local people who offer you their warm hospitality and an insight into their traditional ways of life.

Countries: Egypt, Ethiopia, Oman, Jordan

Twitter: @Wind_Sand_Stars

Web: www.windsandstars.co.uk

Facebook: www.facebook.com/windsandstarsjourneys

## Worldwide Experience 17+

Worldwide Experience allows you to work with animals while contributing to global conservation and community programmes. It specialises in conservation projects in Southern Africa but also offers animal rehabilitation projects, veterinary experiences, school groups and game ranger courses. Other activities can include sports coaching and teaching, marine conservation, wildlife photography and filmmaking. All projects are run by specialists in their field who are specifically trained to impart their skills and experience.

Twitter: @WorldwideExp

Web: www.worldwideexperience.com

Facebook: www.facebook.com/WorldwideExperience

## Worldwide Helpers 18+

Worldwide Helpers is a volunteer network that gives volunteers and organisations the power to share and connect. It provides information on low cost volunteer projects. Worldwide Helpers do not arrange placements but eliminate the middleman and put volunteers in direct contact with low cost voluntary organisations throughout the world. There are online blogs and forums and if you are looking for a 'buddy' to volunteer with you can register here.

Countries: Worldwide

Twitter: @WWHvolunteers

Web: www.worldwidehelpers.org

Facebook: www.facebook.com/worldwidehelpers

## Year Out Group 17+

Formed in 1998 to promote the concept and benefits of well structured year out programmes, to promote models of good practice and to help young people and their advisers in selecting suitable and worthwhile projects, Year Out Group is a not-for-profit association of UK-registered organisations that specialise in this field. All the member organisations are carefully vetted on joining and provide annual confirmation that they continue to abide by the Group's Code of Practice and Operating Guidelines.

Countries: **Worldwide**

Twitter: **@YearOutGroupOrg**

Web: **www.yearoutgroup.org**

Facebook: **www.facebook.com/YearOutGroup/**

Please note that inclusion in this list does not imply any sort of recommendation by Cambridge Occupational Analysts. It is your responsibility to research opportunities very carefully and to select those that best match your personal requirements

# Academic, Cultural, Scientific, Technological, Teaching & Sports-related Options

Organisations listed under this heading offer courses lasting anything from a day or two to around nine months. Courses may be academic, vocational or leisure-based. You will normally have to pay tuition fees, together with accommodation and travel costs.

Options here include:

# Art, Design, Drama, Fashion, Music, Performance

## Art History Abroad (AHA) 16+

AHA's 90-day semester course involves travelling from London to France to Italy to study at first hand themes particularly relevant to each place visited. The programme includes two weeks in London, one week in Paris, one week in the south of France, two weeks in Tuscany, then six weeks travelling through Venice, Verona, Florence, Siena, Naples and Rome, together with day excursions to at least six of the following places: Padua, Vicenza, Bologna, Ravenna, Modena, Urbino, Pisa, San Gimignano, Arezzo, Orvieto, Pompeii and Tivoli. You can opt just for the six-week Italian stage or for two-week summer courses.

**Twitter:** @AHAcourses

**Web:** www.arthistoryabroad.com

**Facebook:** www.facebook.com/ahacourses

## Art, Design, Fashion, Performance 16-18

Short courses at Central Saint Martins, University of the Arts in London, are available over weekends or during the Easter, Summer and Christmas holidays. Subjects include Fashion and Textiles, Fine Art, Graphic Design, Media Arts, Theatre and Performance, and Three-Dimensional Design. Fundamental to study at the College are experimentation, innovation, risk-taking, questioning and discovery, within a highly supportive learning environment, no matter which discipline you choose to study.

**Twitter:** @csmshortcourses

**Web:** www.arts.ac.uk/csm/courses/short-courses

**Facebook:** www.facebook.com/CSMshortcourses

## *European Union Youth Orchestra 14-24

The Orchestra is made up of players representing all the member countries of the European Union (EU). The players are selected each year from over 4,000 candidates aged up to 24, who take part in auditions throughout the EU. Once the members have been selected for the year, you are invited to join the Orchestra to rehearse and perform major works on international stages all over the world.

**Twitter:** @EUYOtweets

**Web:** www.euyo.eu

**Facebook:** www.facebook.com/TheEUYO

*This option will no longer be available to UK citizens once the process of UK exit from the EU is complete.

## John Hall Venice 16+

John Hall offers a nine-week pre-university introduction to some of the most thought-provoking achievements in the Western world, from the classical past to today. A week in London introduces themes which are developed in Venice (6 weeks). Optionally, this is followed by a week in Florence and a week in Rome. The course is conducted through on-site visits and a series of lectures by a team of world-class experts; it includes not only art, but also music, world cinema and literature. There are also practical classes offered in studio life drawing and portraiture, photography, cookery and Italian language.

Twitter: @johnhallvenice

Web: www.johnhallvenice.com

Facebook: www.facebook.com/johnhallvenice

## KLC School of Design 18+

KLC offers a range of short courses that provide a concise introduction to the world of interior or garden design. Whether purely for personal enjoyment, to gain a better appreciation of the subject for your own home or garden project, or as a platform for further study, these practical courses provide a compelling and comprehensive introduction. Options include one- and two-day 'toe in the water' workshops, one- and two-week 'immerse yourself' programmes, or one day a week for ten weeks 'savour the enjoyment' courses.

Twitter: @KLCSchool

Web: www.klc.co.uk

Facebook: www.facebook.com/KLCSchoolofDesign

## London College of Communication 14-18

London College of Communication has designed a programme of short courses and residential summer schools, allowing students to experiment with new techniques, build confidence, acquire new skills and be inspired to learn more.

Twitter: @LCCShortCourse

Web: www.arts.ac.uk/lcc/courses/short-courses/summer-school

Facebook: www.facebook.com/londoncollegeofcommunication

## National Association of Youth Theatres 14 to adult

The Association supports the development of youth theatre activity through training, advocacy, participation programmes and information services. It lists training and events and gives details of youth theatres in your area. There are six associate venues working in partnership with NAYT who produce an annual festival for young people. Each festival is a day-long event and is packed with workshops and performances. See the website for information.

Twitter: @nayttweets

Web: www.nayt.org.uk

Facebook: www.facebook.com/nationalassociationofyouththeatres

## National Student Drama Festival 16-25

The National Student Drama Festival features performances, workshops, discussions, masterclasses and entertainment. The Festival is a seven-day celebration of young people's theatre and features the best of British work alongside exceptional productions by young people from across the world. It is for anyone who wants to get involved with drama and the creative industries and is a place where the most inspirational work from diverse, young and emerging artists is celebrated and presented alongside workshops from leading professionals from around the world.

**Twitter:** @nsdfest

**Web:** https://nsdf.org.uk

**Facebook:** www.facebook.com/nsdfest

## National Youth Orchestra of Great Britain 13-19

One of the world's finest youth orchestras, the National Youth Orchestra (NYO) draws together each year 160-plus talented musicians, aged up to 19, from all over the UK. The orchestra meets during the school holidays for intensive two-week periods of coaching and rehearsal with leading professional musicians and some of the world's finest conductors and soloists, before performing in iconic concert halls such as Symphony Hall, Birmingham, The Sage Gateshead and the Royal Albert Hall.

**Twitter:** @NYO_GB

**Web:** www.nyo.org.uk

**Facebook:** www.facebook.com/NYOGB

## National Youth Orchestras of Scotland 8-25

The National Youth Orchestras of Scotland (NYOS) is unique in providing complete orchestral experience for its students aged between 8 and 25, through its comprehensive range of orchestras, jazz bands and training ensembles. NYOS organises training, intensive rehearsals and national and international concert tours. NYOS is committed to introducing musical experiences to all of Scotland's young people.

Twitter: @NYOScotland

Web: www.nyos.co.uk

Facebook: www.facebook.com/nyos.scotland

## Year Out Drama Company 18+

Year Out Drama provides an intensive, practical drama course, led by experts in professionally equipped performance spaces. This is a full-time programme with a theatre company feel. The course includes Acting, Directing, Design, Costume, Voice Work, Movement, Text Study, Theatre Trips and at least four full-scale performances during the year, including a production at the Edinburgh Festival.

Twitter: @YearOutDrama

Web: www.yearoutdrama.co.uk

Facebook: www.facebook.com/YearOutDramaCompany

## Young Film Academy 6-19

Young Film Academy (YFA), a member of the BFI Film Academy network, is linked directly to the film industry, and helps over 7,500 young people each year complete their first digital films. Programmes range from one day to 2-week residential acting and filmmaking courses and focus on the hands-on approach to filmmaking.

Web: www.youngfilmacademy.co.uk

Twitter: @YoungFilm

Facebook: www.facebook.com/youngfilmacademy

## Au Pair work

### Au Pair in America 18+

Working as an au pair in America can be a good way to discover the USA, as you can experience everyday life with a carefully selected American family and earn weekly pocket money by providing childcare as a nanny or babysitter. Whether you're looking for a year out or just want to work legally abroad, the exchange programmes on the Au Pair in America website give you free time to explore, study, travel and make new friends, together with professional support throughout your stay.

**Countries:** USA

**Twitter:** @aupairamericauk

**Web:** www.aupairamerica.co.uk

**Facebook:** www.facebook.com/aupairamerica

## Cookery

### Avenue Cookery School, London 16+

The Avenue Cookery School offers a variety of courses from basic to advanced, including 'The Really Useful and Chalet Course', a five-day course for undergraduates and Gap Year students covering cooking and 'really useful' information, or the 'Off to Uni' course on how to cook nutritious food on a student budget. The relaxed and friendly atmosphere at The Avenue can turn even the most nervous beginner into a confident 'foodie'.

**Twitter:** @AvenueCookery

**Web:** http://theavenuecookeryschool.com

**Facebook:** www.facebook.com/theavenuecookeryschool

## Ballymaloe Cookery School, Ireland 16+

Run by Ireland's most famous TV cook Darina Allen, Ballymaloe offers a highly regarded 12-week certificate course, graduates of which are in demand all over the world. There is also a wide range of shorter courses – some suitable for complete beginners, others aimed at more experienced cooks. A special time every day is lunch, when teachers and students sit down together to enjoy a three course meal, which the students have prepared using recipes from the demonstrations. After school is over for the day, if the weather is good there are plenty of outdoor activities – tennis, swimming, walking or just lounging around in the beautiful gardens.

Twitter: @BallymaloeCS

Web: www.cookingisfun.ie

Facebook: www.facebook.com/BallymaloeCookerySchool

## Cordon Bleu Culinary Skills 18+

Le Cordon Bleu is a world-renowned network of educational institutions dedicated to providing the highest level of culinary and hospitality instruction through world-class programmes. You might try the four-week 'Essentials Course' at the institute's London School, or you could consider a short course at one of its centres in France, Canada, Japan, Australia, Mexico or South Korea. Le Cordon Bleu is considered to be the guardian of French culinary technique through its culinary programmes that continue to preserve and pass on the mastery and appreciation of the culinary arts that have been the cornerstone of French gastronomy for over 500 years.

Twitter: @LeCordonBleuLDN

Web: www.cordonbleu.edu

Facebook: www.facebook.com/LeCordonBleuLDN

## Edinburgh School of Food and Wine 16+

Among a wide range of courses, you may be particularly interested in the Edinburgh School's four-week intensive certificate course, designed to help you earn your keep during a gap year. This practical course will give you the fundamental skills needed to cook for, say, a ski chalet or a highland lodge, and a grounding in cookery for life. The six-month Diploma in Food and Wine is the School's flagship course, popular with school leavers looking to learn about the culinary world, not to mention the opportunity of job prospects at the end of the programme.

Twitter: @ESFW

Web: www.esfw.com

Facebook: https://www.facebook.com/ESFW1

## Leiths School of Food and Wine, London 13+

This leading UK cooking school offers a range of courses for professional cooks and enthusiast amateurs. 'One Week Cookery Bootcamp' offers a fun-packed introduction to cooking and is perfect for those with a passion for food and entertaining and ideal for those heading to sixth form or university. The course is filled with practical tips and techniques for the budding chef, from baking the ultimate brownies to the perfect Sunday roast. Other courses include 'Let's Cook! Cookery Toolbox for Teenagers' for 13- to 17-year-olds and a range of one-day teenager classes in the summer.

Twitter: @Leithscooking

Web: www.leiths.com

Facebook: www.facebook.com/leithscook

## Nick Nairn Cook School, Scotland 16+

TV chef Nick Nairn is the man behind the Nick Nairn Cook School, a foodie haven built around a single objective: teaching kitchen confidence. The cook school offers a variety of courses where visitors can pick up new skills, inspiration and knowledge whilst preparing their own gourmet lunch. There are two schools: one hidden away in the foothills of the Trossachs, right at the heart of some of Scotland's finest scenery, and the other, their new urban cook school, located in Aberdeen.

Twitter: @NickNairn

Web: www.nicknairncookschool.com

Facebook: www.facebook.com/nicknairnchef

## Orchards Cookery, Worcestershire 16+

This award-winning cookery school won Cookery School of the Year in 2013 and was also Best English Cookery School. The one- and two-week 'Chalet Cooks' courses can show you how to master the art of being a great chalet cook, enabling you to get the most out of a gap year job in the mountains. The school also offers a recruitment service, helping students with CV writing, interview training and so on. Alternatively, the five-day 'Off to University' course covers healthy, delicious and affordable meals for students, including easy entertaining.

Twitter: @OrchardsCookery

Web: www.orchardscookery.co.uk

Facebook: www.facebook.com/OrchardsCookery

## Rick Stein's Seafood School, Padstow, Cornwall 15-16

TV chef Rick Stein's famous cooking school mostly caters for adults, but does offer half-day courses for students during the Easter and summer holidays. For both those who've never cooked with fish and seafood enthusiasts alike, the half-day course teaches you the techniques you'll need in order to create really impressive fish dishes.

**Twitter:** @TheSeafood

**Web:** www.rickstein.com/school

**Facebook:** www.facebook.com/theseafood

## Tante Marie Culinary Academy, Surrey 16+

Renowned for its professional Diploma courses, Tante Marie offers three shorter courses particularly suited to readers of this book. Whether you are keen to go for a gap year job in a stunning location, would like to take your culinary skills to a higher level or just want a short introduction to cooking well for yourself, family and friends, there could be a suitable course for you. The one-term 'Cordon Bleu Certificate' is highly valued by ski companies and other gap year employers, although you might also consider the four-week 'Essential Skills' course. If you simply want to eat well at university, look at the one-, two- or four-week 'Food Camp' courses, offering an introduction to good food and healthy eating.

**Twitter:** @TanteMarie

**Web:** www.tantemarie.co.uk

**Facebook:** www.facebook.com/TanteMarieCulinaryAcademy

# Journalism

## Brighton Journalist Works 18+

Brighton Journalist Works, in partnership with Brighton's daily newspaper 'The Argus', offer a 16-week NCTJ fast-track Diploma in Journalism. The NCTJ qualification is your stepping stone to a job in the industry and is what most UK editors require you to have. The course trains you for newspapers, magazines and online, and prepares you for jobs as a reporter and as a sub-editor. It is a multi-media journalism course which teaches you everything you need to become a journalist – researching and writing stories and features, sub-editing, video production and shorthand, as well as the essential law and public affairs modules. Students have access to unpublicised jobs and internships, and are guaranteed three weeks' work experience. You will learn from working journalists in a friendly, fun and professional environment.

**Twitter:** **@journalistworks**

**Web:** **www.journalistworks.co.uk**

**Facebook:** **www.facebook.com/JournalistWorks**

## News Associates 18+

Based in London and Manchester, News Associates offer NCTJ-accredited courses in Journalism. The 20-week 'Fast Track' course for graduates covers reporting, portfolio, shorthand, law, public affairs and either sport or sub-editing. With a packed timetable and 40 hour weeks, it is not for the faint-hearted! News Associates holds free monthly workshops when those interested in journalism training can try simulated news exercises. There is an intensive two-week summer school for younger students, covering the basics of journalism, news, magazine and feature writing, with an added emphasis on the growing importance of broadcast skills.

Twitter: @newsassociates

Web: http://newsassociates.co.uk

Facebook: www.facebook.com/newsassociatestraining

## Press Association Training 18+

Press Association Training offer NCTJ-accredited courses in multimedia journalism. The courses are based in London, at the Press Association's headquarters, or in Newcastle in the offices of Trinity Mirror. The 17-week full-time course runs twice a year, in spring and autumn. You'll learn all the basics of being a great reporter and have the chance to produce stories for the 'Evening Chronicle', 'The Journal' and the 'Sunday Sun' on the Newcastle course, and various London newspapers if you are studying in the capital.

Twitter: @pa_training

Web: www.becomeajournalist.co.uk

Facebook: www.facebook.com/press.associationtraining

## Languages/Internships

### Bridging The Gap China 18+

Bridging The Gap China offers courses based in the Yunnan province of China that combine Mandarin learning with sightseeing and cultural activities. The courses provide a unique Mandarin learning experience with emphasis on learning whilst travelling. Consequently, they limit the amount of classroom time and ensure a teacher travels with students everywhere. Not only will you get the chance to practise what you've learned in real life situations, but you will also visit some of China's most spectacular sights.

**Countries:** China

**Twitter:** @BTG_China

**Web:** www.bridgingthegapchina.co.uk

**Facebook:** www.facebook.com/Bridging-the-Gap-China-164286446997970

## Cactus Language — All ages

Cactus Language help over 15,000 people every year learn some 25 languages in 120 locations, based on the idea that learning a language in the country where it's spoken is a fun, effective and confidence-boosting experience. They provide thousands of different courses around the world and can create or tailor any programme specifically to individual needs. Cactus also provide courses for children, under-18s, families and the over-50s. There is the opportunity to learn Spanish in Buenos Aires or Italian in Florence. They also offer TEFL courses for those who wish to teach English abroad.

**Countries:** Worldwide

**Twitter:** @cactuslanguage

**Web:** www.cactuslanguage.com

**Facebook:** https://en-gb.facebook.com/cactuslanguagetraining

## CESA Languages Abroad — 16+

If you want to learn Spanish in Spain or Latin America, attend a French course in France or Guadeloupe, study Italian in Italy, or do a German course in Germany or Austria, CESA have a course to suit you! You can learn Portuguese in Portugal, Japanese in Japan, Russian in Russia, Chinese in China or Arabic in Morocco. Whether you are looking for a Summer School course, are planning a Gap Year, want Exam preparation courses or are looking to combine learning a language and activities, there's a wealth of language courses abroad to suit all ages and abilities.

**Countries:** Worldwide

**Twitter:** @CESALanguages

**Web:** https://cesalanguages.com

**Facebook:** www.facebook.com/cesalanguages

## don Quijote 18+

Travel to Spain or Latin America with don Quijote and learn Spanish at one of their friendly schools where you can combine Spanish language lessons with a wide variety of activities, from dance practice to cooking classes. Don Quijote schools are a great place to meet new people. As well as gaining an invaluable knowledge of the Spanish language, you will inevitably come away with a new group of friends from all over the world.

Countries: **Argentina, Bolivia, Chile, Costa Rica, Cuba, Dominican Republic, Ecuador, Guatemala, Mexico, Peru, Spain**

Twitter: @don_Quijote

Web: www.donquijote.org

Facebook: www.facebook.com/donQuijoteSchools

## ESL - Language Travel 18+

ESL offer language courses in over 250 locations around the world. Whatever your level, whether you want to learn Spanish, French, Italian, German, Chinese or any of the many other languages on offer, ESL can assist with everything from accommodation and flights to visa applications and course enrolment. With ESL's Gap Year programme, you can learn a language to an excellent standard while fully immersed in a new culture. The programmes last from three months and are tailor-made, according to your needs and objectives, by their experienced language travel consultants. ESL can also arrange voluntary work, internships and paid work in your destination.

Countries: Worldwide

Twitter: @esleducation

Web: www.esl.co.uk

Facebook: www.facebook.com/ESLlanguagestudiesabroad

## Hutong School 16+

Hutong School is a Chinese Language School with locations in Beijing, Shanghai, Chengdu, London, Brussels, Milan, Moscow, Sydney and Paris. It specialises in Chinese Language Training and Internship Programmes in China, in Beijing, Chengdu and Shanghai. The Chinese classes are characterised by their small group size, individual attention to students' needs and highly qualified and motivated teachers. For the Internship Programme the school cooperates with hundreds of partner companies who are looking for talented students and young professionals to join their teams for short- or long-term internships in China.

Countries: China

Twitter: @hutongschool

Web: www.hutong-school.com

Facebook: www.facebook.com/HutongSchool

## Karstaway 18+

Karstaway offers a structured, 12-week, three-part programme in China, starting with intensive Mandarin lessons in the village of Yangshuo; part two is an internship which involves you taking part in a community-based sustainable development project, while part three gives you the chance to meet all your fellow volunteers back in Yangshuo to reflect on your projects and acknowledge your achievements.

Countries: China

Web: www.karstaway.com

Facebook: www.facebook.com/karstawayadventures

## Language Courses Abroad — All ages

If you are interested in learning Spanish in Spain, studying French in France, doing an Italian course in Italy, German course in Germany, or want to learn Portuguese, Russian, Arabic, Chinese, Japanese or Greek, Language Courses Abroad offer a course and location which could suit your needs and interests. They also offer combination courses, such as skiing and snowboarding in the snow-capped mountains of Europe, Canada or South America with French, German or Spanish language. Or you can give back to the local community whilst totally immersing yourself in the language and culture of the region on a language programme combined with volunteer work.

Countries: Worldwide

Web: www.languagesabroad.co.uk

## OISE — 7-17

OISE language coaching specialises in intensive language courses around the world: English in the UK, USA and Canada, French in Paris and Montreal, Spanish in Madrid and Segovia, German in Heidelberg.

Countries: Canada, UK, USA, France, Spain, Germany

Twitter: @OISE_English

Web: www.oise.com

Facebook: www.facebook.com/OISElanguagecoaching

## Spark Spanish 17+

Spark Spanish runs dynamic Spanish language courses in the city of El Puerto de Santa María in Andalusia in southern Spain. Experience the local culture and authentic lifestyle that comes with studying in this historic town. Spark also runs Work Experience opportunities where, for a minimum of four weeks, students can use their linguistic, creative and technical abilities to promote and improve all that the school has to offer, all with free accommodation included. Spark Spanish also provides a Demi-Pair programme, which gives students the opportunity to stay with a Spanish family whilst gaining experience of looking after children. These opportunities allow students to experience the Spanish way of life whilst simultaneously improving their language skills.

Countries: Spain

Twitter: @SparkSpanish

Web: http://sparkspanish.com

Facebook: www.facebook.com/SparkSpanish

# Medicine, Veterinary Science

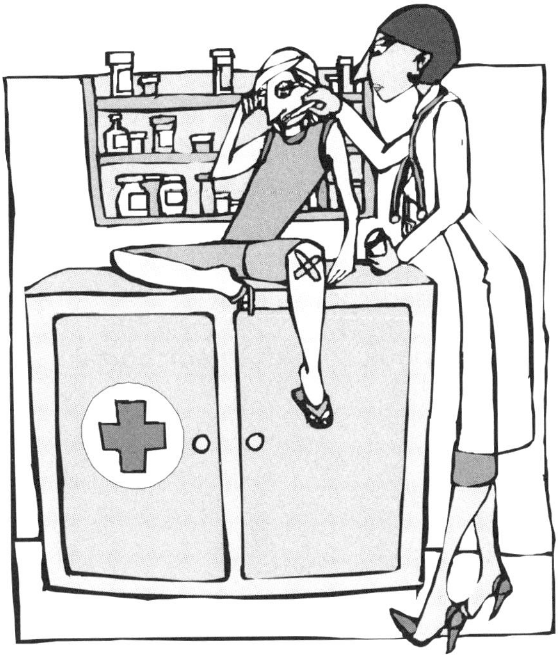

## Gap Medics 16-25

Gap Medics specialises in medical work experience projects. Its one- to eight-week placements offer aspiring medics an invaluable insight into working in a hospital environment. Students spend at least 23 hours per week shadowing a doctor, dentist, nurse or midwife. They also have the opportunity to attend global health tutorials, where doctors from local hospitals provide seminars on local and national health issues. All students stay in custom-built Gap Medics houses. Evenings and weekends are filled with activities such as safaris, hill tribe treks and social barbecues.

Countries: Croatia, Dominican Republic, Poland, Thailand

Twitter: @GapMedics

Web: www.gapmedics.co.uk

Facebook: www.facebook.com/gapmedics

## Medic Portal 16+

Offering a range of courses relevant to medical school application, including planning for the admissions tests, writing the Personal Statement and preparing for interview, Medic Portal also runs a two-day 'How to get into Medical School' conference in partnership with Brighton and Sussex Medical School.

Twitter: @themedicportal

Web: www.themedicportal.com

Facebook: www.facebook.com/themedicportal

## Medlink Campus 16+

Medlink Campus is a three- to five-day course for young people considering a career in medicine. It gives delegates the opportunity to discuss medical school admissions and careers in medicine with Deans from a number of medical schools, as well as advice on surviving medical school given by medical students. It also provides the chance to talk with practising doctors and medical students. The course is valuable for students in years 12 and 13.

Twitter: @medlinkteam

Web: https://medlink-uk.net

Facebook: www.facebook.com/pages/Medlink/162758583877906

## Medsim 16+

Medsim is a two- or three-day residential course held at the Nottingham University School of Medicine. Medsim offers a rich selection of patient contact and practicals that will considerably strengthen your UCAS application. Most importantly, the experience of working under supervision in small groups, with real patients and equipment, has the benefit of allowing you to experience what it is like to be a doctor, to deal with patients, to be on-call and to work under pressure.

Twitter: @theworkshopteam

Web: http://workshop-conferences.com/medsim

## Pre-Med Course 14-18

Pre-Med Course is a one-day medical careers course run by a team of doctors. The course gives impartial careers information to anyone considering medicine as a career. Topics covered include application procedures, interviews, medical school curriculum, and practical demonstrations. Courses are normally held in September, December and April of each year.

Twitter: @PreMedUK

Web: www.premed.org.uk

Facebook: www.facebook.com/premeduk

## Workshop Conferences 16+

Workshop Conferences offer short residential courses, covering such career/degree-related topics as: nursing, medicine, physiotherapy, psychology, veterinary science, chemistry, physics, forensics, journalism, dentistry and law.

Twitter: @theworkshopteam

Web: http://workshop-conferences.com

# Office and IT Skills

## Oxford Media and Business School 17+

The School's one-year Executive PA diploma focuses on the core skills of IT and shorthand but also covers how to deliver powerful presentations and how to research and write persuasive business and marketing reports. Some students view the course as an alternative to higher education, while others see it as a guarantee of well-paid employment before, during and after university.

Twitter: @oxfordbusiness

Web: www.oxfordbusiness.co.uk

Facebook: www.facebook.com/OxfordMediaAndBusinessSchool

## Pitman Training 17+

Pitman training offer flexible, self-paced training, which is tailor-made to fit around current work commitments and lifestyle. Courses are available in areas such as IT and Business skills, including Bookkeeping and Accounts, Web Design, Shorthand, Spreadsheets and Word Processing.

Countries: **Cayman Islands, Ghana, Kenya, Kuwait, Ireland, Nigeria, Pakistan, Palestine, Romania, Russia, Spain, UK, Zimbabwe**

Twitter: **@PitmanTraining**

Web: **www.pitman-training.com**

Facebook: **www.facebook.com/pitmantraininghq**

## Quest Professional 16+

Quest Professional offers practical business skills training to school leavers and university graduates looking to get the skills that employers want and fill in the gaps on their CV. You will develop your practical office IT skills and business communication, as well as increasing your commercial awareness and knowledge of marketing, management, business finance, social and digital media. Throughout the training programmes, you will focus on your future career, CV, job hunting and interview skills technique.

Twitter: **@QuestLondon**

Web: **www.questprofessional.co.uk**

Facebook: **www.facebook.com/quest.london**

# Science, Technology, Engineering, Maths (STEM)

## British Science Association 5-19

The British Science Association organises British Science Week every March, with over 4,500 scientific, engineering and technology events occurring throughout the country, and the British Science Festival in September, which includes dialogue events for 14- to 19-year-olds. The Association also runs the nationally recognised CREST Awards, a project-based award scheme for STEM subjects. Information is sent direct to schools, and more details can be found on the website.

**Twitter:** @BritSciAssoc

**Web:** www.britishscienceassociation.org

**Facebook:** www.facebook.com/BritishScienceAssociation

## Headstart Courses 16+

Headstart is a well-established education programme whose aim is to encourage students interested in mathematics, science or engineering to consider technology-based careers. It provides an opportunity for you in Year 12/S5 to spend up to a week at university prior to making your UCAS application.

Twitter: @TheEDTUK

Web: www.etrust.org.uk/headstart

Facebook: www.facebook.com/TheEDTUK

## Royal Institution All ages

An important part of the work of the Royal Institution is to promote an understanding of science in young people. To further this aim, lectures, events and summer schools are held specifically targeted at the new generation of budding scientists.

Twitter: @ri_science

Web: www.rigb.org

Facebook: www.facebook.com/royalinstitution

## Smallpeice Trust 12-17

The Smallpeice Trust is an independent charity that works to encourage more young people to consider a career in Science, Technology, Engineering and Maths. As part of this, the Trust offers three- to five-day residential courses to provide students with a real-life insight into various strands of engineering. Students in Years 8-12 are offered the opportunity to work with leading engineering companies at top universities around the country, in areas such as Biochemical Engineering, Nuclear Engineering and Physics in Engineering.

**Twitter:** @SmallpeiceTrust

**Web:** www.smallpeicetrust.org.uk

**Facebook:** www.facebook.com/TheSmallpeiceTrust

## Workshop Conferences 16+

Workshop Conferences offer short residential courses, covering such career/degree-related topics as: nursing, medicine, physiotherapy, psychology, veterinary science, chemistry, physics, forensics, journalism, dentistry and law.

**Twitter:** @theworkshopteam

**Web:** http://workshop-conferences.com

## Sports/Outdoor-related

### Alltracks Academy 18+

Alltracks Academy is a family-run company that provides ski and snowboard courses in Canada. Alltracks courses include early morning powder runs, racing gates, recognised instructor qualifications, avalanche safety training, and backcountry adventures. Held in the world class ski resort of Whistler, Alltracks courses range from a couple of weeks to the whole season. They make an ideal way for keen skiers or snowboarders to spend a constructive gap year or career break in the mountains, or help kick-start an instructor career.

**Countries:** Canada

**Twitter:** @ALLTRACKS

**Web:** www.alltracksacademy.com

**Facebook:** www.facebook.com/ALLTRACKS

## Altitude Futures 17+

Altitude Futures offer you the chance to become a fully qualified ski or snowboard instructor. Aimed at competent skiers and snowboarders, the 10-week gap instructor training course provides you with the opportunity to gain a level 1 and 2 BASI (British Association of Snowsport Instructors) ski or snowboard instructor licence, first aid qualification and practical teaching experience whilst skiing or snowboarding at one of the top resorts in the world in Verbier, Switzerland. Each year, Altitude guarantee 10 jobs to the best students.

**Countries:** Switzerland

**Twitter:** @AltitudeVerbier

**Web:** www.altitude-futures.com

**Facebook:** www.facebook.com/altitudeverbier

## AmeriCamp 18+

AmeriCamp offers people around the world the chance to work at a summer camp in America, either as a camp counsellor or in a support role behind the scenes. Each year AmeriCamp holds camp fairs in Dublin, Glasgow, Leeds and London, which give applicants a chance of meeting camp staff face-to-face and securing a job on the day. It's also a great way of finding the right camp for each person.

**Countries:** USA

**Twitter:** @AmeriCamp

**Web:** www.americamp.co.uk

**Facebook:** www.facebook.com/AmeriCamp

## Camp America 18+

Each year over 7,500 young people take the opportunity to join Camp America and spend the summer in the USA, living and working either with children or behind the scenes as support staff on an American Summer Camp. Following the end of your placement, you'll have up to two months to travel (in total your visa allows for up to four months placement followed by one month of travel but most placements end well within four months).

Countries: USA

Twitter: @CampAmerica69

Web: www.campamerica.co.uk

Facebook: www.facebook.com/campamerica69

## Flying Fish 17+

Flying Fish trains and recruits over 1,000 people each year to work as yacht skippers and as sailing, diving, surfing, windsurfing and ski and snowboard instructors. As an experienced yacht skipper or instructor, you can then spend time earning money from your favourite sport. Courses range from one week to 19 weeks and are available to complete beginners and for those who already have some experience.

Twitter: @Flyingfish_sail

Web: www.flyingfishonline.com

Facebook: www.facebook.com/flyingfishonline

## Jonathan Markson Tennis 10+

A graduate of Christ Church College, Oxford, Jonathan Markson is a former captain and coach of the Oxford University 'blues' tennis team and an international player for Scotland. His company offers tennis holidays and tennis camps in England, Portugal, Spain, Italy, South Africa, Canary Islands and Mauritius. The camps aim to provide players of all ages and levels with a rewarding and enjoyable tennis experience, where the coaching is serious but fun.

Twitter: @Marksontennis

Web: www.marksontennis.com

Facebook: www.facebook.com/Marksontennis

## Jubilee Sailing Trust 16+

The Jubilee Sailing Trust (JST) is a charity that aims to promote the integration of people of all physical abilities through the challenge and adventure of tall ship sailing. The JST owns and operates two tall ships – LORD NELSON and TENACIOUS – the only two vessels in the world that have been purpose-designed and built to enable a crew of mixed physical abilities to sail side by side on equal terms. If you take on the tall ship challenge with the JST, it could be a short hop around the British coast, a four-week transatlantic challenge, a week's island hopping in the Canary Islands or the Caribbean, or a place in the European Tall Ships' Race.

Twitter: @JubileeSailing

Web: www.jst.org.uk

Facebook: www.facebook.com/JubileeSailingTrust

## Manor House Activity and Development Centre 18+

Manor House offers a 16-week intensive Outdoor Instructor Training course. During your time on the course you will develop your skills in specific activities: climbing, kayaking, sea kayaking, canoeing, stand up paddleboarding, mountain skills and navigation, cycling, beach lifeguard skills and first aid. You will also work on the theoretical and practical aspects of risk assessments, hazard management, group management and leadership, to name but a few of the modules included.

**Twitter:** @MActivityCentre

**Web:** www.manoractivitycentre.co.uk

**Facebook:** https://en-gb.facebook.com/manoractivitycentre

## NONSTOP Adventure 17+

NONSTOP Adventure is a family-owned company that runs sports coaching and skiing, snowboarding and surfing training courses. Whether your passion is for the sea or the mountains, there is a course for you. In most cases courses will result in gaining internationally recognised instructor qualifications that could open up employment opportunities worldwide. All courses are run by the industry's top professionals and focus on general improvement as well as gaining professional instructor qualifications. All abilities welcome.

**Twitter:** @nonstopsnow

**Web:** www.nonstopadventure.com

**Facebook:** https://m.facebook.com/nonstopsnow

## Outward Bound — 11 to 19

The Outward Bound Trust runs adventure programmes ideal for young people looking to do something different with their summer. It offers a range of one-week to 19-day experiences, run throughout the summer months, including a range of specialist climbing, biking or paddling adventures. Each programme not only gives you a chance to try out new activities, from gorge scrambling to expeditions in the mountains, but also equips you with new skills and can make a real difference to your CV or university application. Choose from three locations: Loch Eil in the Scottish Highlands, Ullswater in the Lake District or Aberdovey in Snowdonia, Wales.

Twitter: @OutwardBoundUK

Web: www.outwardbound.org.uk

Facebook: www.facebook.com/outwardbounduk

## Peak Leaders — 17+

Peak Leaders' Mountain Bike, Ski and Snowboard courses ensure you will be improving your technical skills, gaining snowsports qualifications, and increasing your understanding of mountain environments, safety and team leading. At the same time, you'll be experiencing life in another culture. There are also nine-week summer break, southern hemisphere courses, with the emphasis on travel and adventure.

Countries: Argentina, Switzerland

Twitter: @PeakLeaders

Web: www.peakleaders.com

Facebook: www.facebook.com/PeakLeaders

## PGL 18+

PGL provides children's adventure holidays and educational courses at its 23 activity centres across the UK, France and Spain. Every year it recruits over 2,500 staff to instruct, inspire and look after its guests, with vacancies for watersports instructors, adventure activity instructors, group leaders, language speakers, TEFL Teachers, administrators, maintenance, catering and domestic staff. There are also ad hoc ski rep positions for the peak weeks of the winter operating season.

**Countries:** UK, France, Spain

**Twitter:** @pglstaff

**Web:** www.pgl.co.uk/jobs

**Facebook:** www.facebook.com/pglstaff

## Quaystage Training 14-25

Quaystage delivers water-based outdoor activity adventures and training. These provide a platform for experiential learning, personal development and achievement. Opportunities include Sail Camp and Dive Camp adventures during school holidays, and Tall Ship sailing adventures.

**Twitter:** @BritSailTrain

**Web:** www.quaystage.co.uk

**Facebook:** www.facebook.com/QuaystageTraining

## ROCK Sailing — 16+

ROCK Sailing offer professional Yachtmaster training in the sunshine of the Mediterranean. Based in Gibraltar they combine professional training with adventure on courses lasting from one to 14 weeks, designed to take a complete newcomer to sailing through to being a fully qualified RYA Yachtmaster and Sailing Instructor.

Countries: Gibraltar, Spain

Twitter: @RockSailing1

Web: www.sailinggibraltar.co.uk

Facebook: www.facebook.com/pages/Rock-Sailing-Gibraltar/145216602172279

## Snow Academy — 18+

Snow Academy offers ski and snowboard instructor courses, together with chalet host cookery training, in Whistler, Canada. Choose from 4-, 7- or 11-week courses at Whistler's premier ski school and achieve recognised instructor qualifications to teach skiing/snowboarding all over the world.

Countries: Austria, France, Italy, Norway, Switzerland

Twitter: @skiweekends

Web: www.skiweekends.com/ski-weekends/skiweekends-snow-academy

Facebook: www.facebook.com/internationalacademy

## Sporting Opportunities 17+

Sporting Opportunities offer sports coaching volunteer projects which give you the chance to travel to many different parts of the world with like-minded people and volunteer in the community as a sports coach. You will coach sports to children from disadvantaged backgrounds, play sport in the local community and return home with some unforgettable memories. There are a variety of volunteer sports projects to choose from, including football, hockey, rugby, netball and many more. You could find yourself coaching football in India, or perhaps netball and hockey in South Africa.

Countries: **Australia, Canada, China, Costa Rica, Ecuador, Fiji, India, Kenya, South Africa, St Lucia, Thailand**

Twitter: **@SportingOpps**

Web: **www.sportingopportunities.com**

Facebook: **www.facebook.com/sportingopportunities**

## Tall Ships Adventures 12-15, 16-25

The Tall Ships Youth Trust is dedicated to the personal development of young people aged 12 to 25. The fleet includes Tall Ship Stavros S Niarchos (a magnificent 200ft Brig), four 72ft Challenger Yachts and the 62ft Tall Ships Catamaran. They are operated by Tall Ships and work 12 months of the year, both around the UK and abroad, offering Tall Ship Adventure Sailing Holidays. Recognised by the Duke of Edinburgh Award as an activity provider and approved for the Royal Yachting Association Competent Crew qualification.

Countries: **Predominantly UK and European waters but also the Caribbean**

Twitter: **@TallShipsYT**

Web: **http://tallships.org**

Facebook: **www.facebook.com/TallShipsYouthTrust**

## United Through Sport 18+

United Through Sport is a UK-registered charity that supports sport and recreation projects in Developing World countries. It particularly focuses on outreach programmes to deliver sporting opportunities for over 56,000 disadvantaged children. The sports programmes are run in local communities, focusing on boxing, rugby, netball, cricket, soccer, hockey, swimming and tennis, and offer the opportunity to volunteer as a sports coach.

Countries: Argentina, Caribbean, Ghana, South Africa, Thailand

Twitter: @utscharity

Web: www.unitedthroughsport.org

Facebook: www.facebook.com/utscharity

## Woodlarks Campsite Trust 16+

Situated in twelve acres of beautiful Surrey countryside, Woodlarks Campsite enables children and adults with disabilities to enjoy a host of activities they may never have thought possible. Woodlarks camps can be as tranquil or as adventurous as you want them to be, and you can volunteer as a helper during the camping season.

Twitter: @WoodlarksCamp

Web: www.woodlarks.org.uk

Facebook: www.facebook.com/wcampsite

## Yamnuska Mountain Adventures 17+

Yamnuska Mountain Adventures is a mountain skills and experience organisation. Courses are led by experienced mountain guides and instructors. The company specialises in Mountaineering, Rock Climbing, Ice Climbing, Ski Touring, Trekking and Avalanche education at beginner to expert level. The three-month Mountain Skills and Leadership Semester based in Alberta, Canada gives participants the skills and experiences for safe mountain travel and leadership. Yamnuska arranges your accommodation for the whole trip and provides all meals when you are out in the field.

Countries: Canada

Twitter: @Yamnuska_Mt_Adv

Web: https://yamnuska.com/climbing-school

Facebook: www.facebook.com/yamnuska.mountain.adventures

# Summer Schools

## Cambridge Scholars' Programme (CSP)

CSP is a private Summer School offering students from the UK, US and the rest of the world the chance to live in student accommodation in Cambridge while following an academic study programme combined with visits to sites of interest in Britain and Paris.

The three-week programme allows you to choose two or three topics to study – ranging from Criminal and Forensic Psychology to Cosmology, Scriptwriting or the Quantum Universe. Activities are run after classes and in the evenings.

**Web:** http://cambridgescholarsprogramme.com

**Facebook:** www.facebook.com/CambridgeScholarsProgramme

## Debate Chamber Summer Schools 15-18

Based in London, these Summer Schools offer students the opportunity to explore some fascinating subjects, prepare for university application and experience a challenging approach to learning. Courses are offered in Law, Economics, Politics, Philosophy, History, International Relations, English Literature, Creative Writing, Art History, Medicine, Dentistry, Maths and Physics.

Working in small groups gives you a real chance to get to know your tutors and fellow students. The Schools are about more than just lectures and seminars. For example, English includes poetry composition workshops, play readings and tickets to a performance at Shakespeare's Globe. Law includes a visit to the Inns of Court and Royal Courts of Justice as well as a full Mock Trial in which students take on the role of Crown Court barristers.

**Twitter:** @debatechamber

**Web:** https://debatechamber.com/summerschools

**Facebook:** www.facebook.com/debate.chamber

## Sutton Trust Summer Schools 16-18

The Sutton Trust Summer Schools are free, subject-specific residential courses for Year 12 students from UK state-maintained schools. The summer schools allow bright students from non-privileged homes a taste of life at a leading university and an insight into what it is like to live and study as a first-year undergraduate student. The summer schools balance busy academic days with enjoyable social activities. The aim is to demystify elite universities and to equip students – most of whom will be the first in their families to go on to higher education – with the knowledge and insight to make high quality applications to top universities.

**Twitter:** @suttontrust

**Web:** https://summerschools.suttontrust.com

**Facebook:** www.facebook.com/suttontrust

## Working Holiday/Cultural Exchange

### Australia Working Holiday 18-30

If you are aged between 18 and 30 and hold a passport for a country or region participating in Australia's Working Holiday Maker programme, you may be eligible to apply for a 12-month visa which enables you to work in Australia. The Australian Government's Working Holiday Maker Programme is a cultural programme which enables young travellers to have an extended holiday and earn money through short-term employment. If you work in Australia's regional areas, in certain circumstances you may even be able to extend your stay for an additional 12 months.

Countries: Australia

Twitter: @AusBorderForce

Web: www.homeaffairs.gov.au/Trav/Visi/Visi-1

Facebook: www.facebook.com/aushomeaffairs and www.facebook.com/AustralianBorderForce

## Kibbutz Volunteers 18-35

Living and working in a kibbutz community in Israel, carrying out the true principles of a socialistic society, having all work, property and profit equally shared by its members, can form the basis of an intriguing working holiday experience. A holiday with the possibility to meet, live and work with both Israeli youngsters and other kibbutz volunteers from countries and cultures far and near.

Countries: Israel

Twitter: @KIBBUTZwebsit

Web: www.kibbutz.org.il/eng

Facebook: www.facebook.com/kibbutzvol

## IST Plus 17+

With an IST Plus cultural exchange programme you can work, study, travel or teach in locations throughout the world. Work in the USA, teach in Thailand or China, travel around Australia, New Zealand and Singapore, or study a language.

Countries: Australia, Cambodia, China, New Zealand, Singapore, Thailand, USA

Twitter: @ISTPlusWorld

Web: www.istplus.com

Facebook: www.facebook.com/istplus

## Rotary Youth Exchange 15-19

Each year the Rotary Youth Exchange programme sends literally thousands of young people, on long- and short-term exchanges, special interest camps and tours. These aim to promote an insight into another country's way of life, traditions and culture, and to develop lasting friendships.

**Twitter :** @RotaryExchange

**Web:** www.youthexchange.org.uk

**Facebook :** www.facebook.com/RYEGBandI

## Visit Oz 17-30

If you are considering a gap year in Australia before you go to university, after you have graduated or at any time before your 31st birthday, Visit Oz guarantees to find you a job on the land or in rural hospitality, as well as providing agricultural or hospitality training. You must have a Working Holiday Visa (or other Visa allowing work) and be prepared to get your hands dirty. Outback farm or station work can include working with horses, cattle and sheep, tractor and header driving, bulldozer work, fencing, mechanical work, and chainsaw work; horse work may be at stables, in trail riding centres, on Host Farms, with racehorses, polo ponies, camp draft horses, or on cattle properties doing bore running, yard work, and maintenance; agricultural bike work is with cattle and sheep. There are so many jobs that it is possible to find something to suit the skills of everybody.

**Countries:** Australia

**Web:** www.australianworkingadventures.com

**Facebook:** www.facebook.com/Visitoz

## Worldwide Opportunities on Organic Farms 17+

A network of organisations covering some 100 countries, WWOOF arranges volunteer work on organic farms across the world, from Argentina to Zimbabwe. In return for volunteer help and a small fee, food, accommodation and opportunities to learn about organic lifestyles are provided. Usually you live with your host and are expected to join in and cooperate with the day-to-day activities. In most countries the exchange is based on 4-6 hours help-fair exchange for a full day's food and accommodation. You may be asked to help with a variety of tasks like sowing seed, making compost, gardening, planting, cutting wood, weeding, harvesting, packing, milking, feeding, fencing, making mud-bricks, wine making, cheese making and bread making. Most WWOOF visits are between one and two weeks, though some may be as short as two or three days or as long as six months.

Twitter: @WWOOF

Web: www.wwoof.net

Facebook: www.facebook.com/WWOOF

# Important Information and Advice

# Fundraising

We mention at several points in this publication that it will cost you a fairly considerable sum – often several thousand pounds – to participate in some of the projects listed. This is particularly true of many of the international community, environmental or scientific projects.

For example, to have gone overseas for 12 months with Project Trust in 2017, you would have had to raise £6,200. Project Trust will raise another £600 on your behalf to subsidise the full costs of your year abroad.

Organisations such as Project Trust receive no government assistance and all funds must be raised either by project managers or by volunteers like you.

Given that the aim of this book is to provide you with ideas to help develop your personal, learning and thinking skills, we believe that raising sponsorship can be an important part of this process. It shows others your determination and initiative, and it will help you establish in your own mind just how well you can respond to a challenge. You will have to start by learning how to fundraise and how to make the most of the support available.

Experienced organisers of such projects say that most volunteers are surprised by the response to their efforts and many not only hit their target but actually raise more than the sum required. Only a small number each year have problems and even they can usually be helped to find suitable sponsors.

Should you decide to opt for a project with a sizeable participation fee, you will find that the organisers will normally send you, once accepted, a comprehensive pack containing fundraising ideas and information. In addition, there should be an experienced member of staff able to give you help and advice by email or over the telephone.

The list below should give you a clear idea of the level of fundraising support you should look for when researching a possible project:

- Advice on fundraising: Does the selection process introduce the idea of fundraising through a seminar or workshop, encouraging you to think about it constructively?
- Ongoing support: What mechanism exists for you to keep the organisers informed of your progress? If you are struggling, do they provide practical advice?
- Fundraising meetings: Will you be invited to one or more fundraising meetings, where you can get together with fellow volunteers to share ideas and experiences?
- Bulletin board: Is there a website bulletin board allowing you and your fellow volunteers to keep in touch?

## Fundraising ideas

The best starting point is always to look inside yourself! There should be no need to turn your life around completely to raise the required funds. Consider what you are already good at and love doing, then think about how you can use your skills to make the money you will need.

If you are good at music, for example, you could try your hand at busking, performing at various events or offering home tuition.

If your interests are more sporting, you could arrange a tournament where teams pay to enter and you provide the service of organising it and setting up suitable prizes.

If you are green-fingered, you could offer a gardening service – weeding, cutting lawns or hedges – or bring on seeds and cuttings to sell at every opportunity.

If you can cook, you could offer a catering service for parties and other social gatherings.

If none of these applies, you could simply get a part-time job of any sort and start saving regularly to establish your fund.

Once you have a service to offer, goods to sell or a job to find, turn first to your immediate circle of family, friends and school, college or other social contacts, perhaps in a youth or sports club. If they can't offer you direct support, ask them to think about who they could put you in touch with, or who they might be prepared to approach on your behalf. Before long, you should have a long list of potential customers/employers/sponsors!

While you are raising money, don't forget that you will have other things to buy. You might, for example, need a top quality sleeping bag or rucksack, both of which are likely to be expensive. Baggage insurance is another important extra. You will also need to pay for your own visa, which could cost up to £250, your medical examination and any necessary inoculations.

When you go, you will need to take some spending money with you: perhaps around £1,000 for a year-long project, although this can depend upon which country you go to. Even if you can afford it, you shouldn't think of taking so much money overseas that you might be tempted to live and travel in a way that would not sit easily with your role as a volunteer.

# Managing Risk

Many of the suggestions in this book contain an element of risk. That is part of their attraction and you will no doubt see little point in trekking through a jungle or across a mountain range if you are going to be as cosy and safe all the time as you are in an armchair at home. Nevertheless, we could not possibly encourage you to take unnecessary risks and we recommend that you venture overseas only with a reputable organisation with experienced leaders and stringent operating procedures designed to avoid foolhardy misadventure.

We cover a broad range of health and safety issues in our brief quiz on page 115 and we would ask you to spend some time reading this section and visiting the recommended websites. Amongst them, the Know before you go site managed by the Foreign and Commonwealth Office is absolutely essential (see page 120).

In addition, we suggest that you use the checklist below when choosing an organisation with which to undertake your trip. This will help to ensure that you are in safe hands and will be travelling responsibly.

- Crisis Management: Is there a comprehensive crisis management policy in place? How robust is it and how are staff trained to implement it?
- UK Support: Does the organisation maintain a 24-hour emergency telephone line for family and friends in the UK?
- Insurance: Does the organisation have a comprehensive company insurance policy with a specialist provider?
- Leaders: What is the organisation's recruitment policy in relation to the experience and qualifications of expedition leaders? What knowledge do they have of the countries they work in? Do they have relevant language and first aid skills? Does the Criminal Records Bureau carry out checks of their backgrounds?
- Risk Assessment: Do expedition leaders undertake daily monitoring of activities in order to maintain the safety of participants? Are written risk assessments available for consultation? Are participants encouraged to carry out their own risk assessments during an expedition?
- In-Country Support: Do leaders have up-to-date contact lists for medical and logistical support in the country you will be visiting?
- Participant Preparation: What level of pre-departure training and/or in-country orientation is provided for participants?
- Equipment: How often is equipment reviewed and replaced? Is safety equipment provided as standard? Do leaders carry comprehensive first aid kits?

- Transport: How do leaders assess in-country transport? Is there a policy regarding the use of public or private transport options? Is road transport undertaken at night?
- Responsible Travel: What is the organisation's policy in relation to monitoring and minimising the long-term impact of expeditions such as the one you are considering? Is there a long-term commitment to cultural sensitivity and sustainable development?
- Financial Transparency: Can the organisation demonstrate that your financial contributions are spent directly on the project and nowhere else?
- Feedback: Is there evidence that feedback from participants and staff is assessed and acted upon where necessary to improve future provision?

A final word. While we have tried our best, as publishers of this book, to ensure that you understand the nature of potential hazards overseas – how to recognise and overcome them – we must stress that you cannot rely totally on us or even on the very best provider of specialist gap year activities. You must use your own common sense and initiative to help you to spend your time as safely as possible.

# Finding a Focus

When you have researched the ideas in this book, photocopy these pages and complete a worksheet for each project that interests you. Our 10-point plan will help you focus on finding the right programme to develop your personal learning and thinking skills.

| Name of organisation |
| --- |
| Type of activity |
| 1. What is it that appeals to me?<br>*(Gaining relevant experience, travel, helping others, earning money)* |
| 2. Am I eligible? *(Right age, available at the right time, suitable qualifications)* |
| 3. What exactly will I be doing?<br>*(Working with others, undertaking/learning about research methods, cookery)* |
| 4. How will I benefit from this programme? |
| 5. How will other people benefit from my involvement? |

**6. How much will it cost?** *(Total budget, raising funds, putting down a deposit)*

**7. Who will I be signing up with?** *(Commercial company, registered charity)*

**8. What do I need to arrange?** *(Travel, insurance, health check, vaccinations)*

**9. Is there any pre-programme training or briefing?**

**10. What happens afterwards?** *(Debriefing, maintain contact with organisations/ providers, inform future participants, obtain certificate recording my achievements)*

## Am I Ready for a Trip Abroad?

If you are planning to travel abroad as part of your personal development, try our brief quiz to see how well prepared you are!

1. INSURANCE ISSUES
   - (a) I will investigate a range of different types of insurance to cover my travel and placement/project ☐
   - (b) I will take out travel insurance for the journeys to and from my placement/project ☐
   - (c) I guess that I'm on our family annual travel insurance and that my parents' insurance policies will cover all eventualities ☐
2. THE COUNTRY I INTEND TO VISIT
   - (a) I have researched the laws and customs of my planned destination as well as the usual food, currency and weather type research ☐
   - (b) I have looked at holiday websites to find out about the country I intend to visit ☐
   - (c) I'll pick up everything I need to know just by being in the country for several months ☐
3. VISAS AND PERMITS
   - (a) I have checked out the necessary visas and work permits for the country I intend to visit ☐
   - (b) I will see if I need a visa ☐
   - (c) Someone will sort out whatever has to happen about a visa for me ☐
4. HEALTH
   - (a) I will check with my local surgery to see if I need any special injections or healthcare for the country I am visiting a couple of months before the departure date ☐
   - (b) I will ask my mum to take a look on the internet to see if there is anything I need to do about health care arrangements for my visit ☐
   - (c) I won't bother to do anything special about healthcare as I am young and healthy ☐
5. SAFETY
   - (a) I have seriously considered a number of ways of ensuring my safety when I am on my placement and I have discussed safety plans with my family ☐
   - (b) I am always careful about my well being and I won't need to do anything extra for my placement ☐
   - (c) I have no safety worries and I can look after myself ☐

## How did you do?

If you answered (a) to all the questions then you have made a good start. All (b) then you have a bit more work to do. All (c) then you really must do a lot more research

## Points to consider

### 1. Insurance issues

It'll never happen to me!

It can happen to you; things can go wrong. You could fall ill or have an accident; you could have money or luggage stolen; your visit might be cancelled or cut short through injury or illness; your family may need to fly out to be with you if there is a serious incident. So take out insurance. Make sure it's comprehensive and covers you for medical and repatriation costs as well as any dangerous sports or activities.

If you get injured or ill as a result of drugs or alcohol, your insurance may be invalidated and your travel operator can refuse to fly you home.

### 2. The country I intend to visit

You must read up on the laws and customs of your chosen destination, to avoid offending people or breaking local laws, however unwittingly. The best starting point for this is the Foreign and Commonwealth Office, with its 'Know before you go' awareness campaign aimed at encouraging British travellers to prepare better before going overseas.

Visit the website at: **https://travelaware.campaign.gov.uk**, and follow them on twitter **@FCOtravel**.

If you are contemplating taking drugs whilst on holiday abroad or bringing some back with you, stop and think – otherwise your trip of a lifetime could end up lasting a lifetime in jail. Bear in mind that: some 6,000 British nationals are arrested overseas each year, a third of them for drugs-related offences, and at any time there are around 3,250 in prisons around the world; many countries outside the UK refuse to grant bail before trial and may detain people in solitary confinement; you will still get a criminal record in the UK if arrested with drugs abroad; if you've been caught with drugs abroad, you're unlikely ever to be allowed to visit that country again.

## 3. Passports, Visas and Work Permits

If you wish to travel abroad you must hold a full ten-year passport, even for a day trip. Apply in good time. In the UK, you can get advice from the HM Passport Office website at **www.gov.uk/government/organisations/hm-passport-office** or call them on the Passport Advice Line on 0300 222 0000 (open 8am to 8pm Mon-Fri and 9am to 5.30pm weekends & public holidays). Some countries have an immigration requirement for a passport to remain valid for a minimum period (usually at least six months) beyond the date of entry to the country. Therefore, if appropriate, ensure your passport is in good condition and valid for at least six months at the date of your return. This is a requirement of the country concerned, not the UK Passport Service, and any questions should be addressed to their Consulate or Embassy.

Outside the UK, you should get advice in an emergency from the nearest British Embassy, High Commission or Consulate. Staff can issue standard replacement passports in most places, and all missions are able to issue emergency passports if more appropriate.

If you plan to travel outside British territories, you may require a visa to enter the country you are going to. Check visa requirements with your project organiser or travel agent or contact the Consulate or Embassy of the country you plan to visit.

If you plan to work outside the European Union, you will need to obtain a valid work permit before you go.

Some Passport Tips:

- Make a note of your passport number, date and place of issue (or take a photocopy), and keep separately in a safe place.
- Check your passport expiry date.
- Write the full details of your next of kin in your passport.
- Leave a photocopy with a friend or relative at home.
- Take a second means of photo-identification with you.
- Keep your passport in the hotel safe and carry a photocopy with you.
- If your passport is lost or stolen overseas, contact the nearest British Embassy, High Commission or Consulate immediately for advice.

## 4. Health

- Check the Department of Health website at: **www.nhs.uk/Healthcareabroad** for general medical advice for travellers.
- Check what vaccinations you need with your GP at least six weeks before you travel.
- Check if your medication is legal in the country that you are visiting.
- Pack all medication in your hand luggage.
- If you are taking prescribed medication, take the prescription and a doctor's letter with you.
- If you are travelling within the European Economic Area (EEA) or Switzerland, you should get a free European Health Insurance Card (EHIC) by visiting the Department of Health website as above. You can also obtain the EHIC by completing the Department of Health leaflet 'Health Advice for Travellers' (HAFT), available through most UK Post Offices or by telephoning 0300 330 1350. The EHIC entitles you to free or reduced-cost medical care but you will still need medical and travel insurance. It is not yet clear what the status of the EHIC will be when the UK formally leaves the European Union (EU), although it is likely that the UK will negotiate similar reciprocal healthcare arrangements with some or all of the remaining EU member states.
- Should you find yourself in an emergency during a visit to Europe, dial 112. The European emergency number is valid in all EU/EEA member states and is free of charge.
- Be safe in the sun. Avoid excessive sunbathing, especially between 11am and 3pm, and wear a high factor sunscreen.
- Drink plenty of water. If you drink alcohol or use some kinds of drugs your body can become dehydrated, especially in a hot climate.
- Find out the local emergency number and the address of the nearest hospital when you arrive overseas. Your rep, local guide or project manager should know.

## 5. Safety

Be aware of what is going on around you and keep away from situations that make you feel uncomfortable. Avoid potentially dangerous 'no-go' areas, in particular after dark. Use your common sense and make sure you are constantly assessing and reassessing your personal safety. Be aware of drugs – these have been used in incidents of rape, so keep your wits about you.

Keep an eye on your possessions. Never leave your luggage unattended or with someone you don't completely trust. Be aware of pickpockets, who tend to operate in crowded areas, and lock up your luggage with padlocks. Make sure you have copies of all important documents such as your passport, tickets, insurance policy, itinerary and contact details. Keep these separate from the originals and leave copies with your family and friends.

Work out how much money you'll need on a daily basis and work to a realistic budget. Be sure to take enough money, as the Foreign and Commonwealth Office can't send you home free of charge if you run out!

Finally, tell friends and family your plans before you go and keep in regular contact, especially if you change your plans. Consider taking a roam-enabled mobile and use text or email to keep in contact. Don't promise too much – promising to call home every day is unrealistic and will only cause your family and friends to worry when you don't!

### Best Gap Year

A site with worldwide gap year jobs, courses and travel opportunities, from winter jobs in a ski resort or summer sports teaching courses in Australia to medical projects in China. It offers a range of ideas across the world with opportunities to take part in conservation and community projects, teaching, sports and volunteering. It also provides advice on gap year planning, health and other issues.

**Twitter:** @bestgapyear

**Web:** www.bestgapyear.co.uk

**Facebook:** www.facebook.com/Best-Gap-Year-104314879685938

### Foreign & Commonwealth Office Website

The FCO website offers advice for anyone thinking of embarking on gap year travel. From top tips on insurance and money to advice on staying healthy and getting the right visa, you will find all you need to know to plan the safest and most enjoyable gap year travel.

**Twitter:** @fcotravel

**Web:** www.gov.uk/guidance/gap-year-foreign-travel-advice

https://travelaware.campaign.gov.uk

**Facebook:** www.facebook.com/foreignoffice (or /fcotravel)

### Gapwork

Gapwork.com works with many of the top gap year companies to help young people find the best gap year options, adventures and placements abroad. The website provides useful news, facts and information on jobs, destinations, activities and volunteering to help cover the whole gap year experience from planning to execution.

**Twitter:** @Gapworkjobs

**Web:** www.gapwork.com

**Facebook:** www.facebook.com/gapwork

## Gapyear.com

Gapyear.com is a social media and travel advice website created by backpackers, for backpackers, devoted to giving you everything you could possibly need when planning or taking a gap year. It hosts and supports a community that is passionate about real travel and ready to share experience and advice. The community is backed up with expert guides.

Twitter: @gapyeardotcom

Web: www.gapyear.com

Facebook: www.facebook.com/gapyeardotcom

## Hostelling International

Youth hostels can provide you with reliable, reasonably priced accommodation in many parts of the world. Hostelling International is the brand name of more than 90 Youth Hostel Associations in 90 countries, operating 4,000 plus hostels. Unlike bland motels, impersonal hotels or dodgy backpacker rooms, youth hostels are usually fun, lively meeting places, full of like-minded people.

Twitter: @Hostelling

Web: www.hihostels.com

Facebook: www.facebook.com/HostellingInternational

## Objective Gap Safety

During Objective's Gap Safety courses, students are given advice, warnings and responses to various situations they may face while travelling. Courses are designed to teach individuals how to evaluate situations to avoid trouble and to give them the confidence to get off the beaten track and explore.

Twitter: @ObjectiveTravel

Web: www.objectivegapyear.com

Facebook: www.facebook.com/ObjectiveGapSafety

## STA Travel

STA Travel are specialists in cheap flights, adventure trips, insurance, bus tickets, rail passes, car/campervan hire, hostels, and travel deals for young people. They also have several pages of gap year travel tips on their website.

Twitter: @STATravel_UK

Web: www.statravel.co.uk

Facebook: www.facebook.com/STATravel

## World Travel Guide

Want a heads up on what the visa regulations are for visitors to India, how much duty free you can bring back from Chile, what the main sights are in Yemen or when is the best time of year to go to Jamaica? It's all available here, and quickly searchable by country or region.

Twitter: @WTGTravelGuide

Web: www.worldtravelguide.net

Facebook: www.facebook.com/WTGTravelGuide

# Suggested Reading List

The Big Trip: your Ultimate Guide to Gap Years and Overseas Adventures
- Lonely Planet, 2015

Gap Year Adventures: an Essential Pocket Guide to Making It Count
- Tamsin King, Summersdale, 2016

Gap Years – The Essential Guide
- Emma Jayne Jones, Need2Know, 2013

Backpacking: The Ultimate Guide to First Time around the World Travel
- Michael Huxley, CreateSpace Independent Publishing Platform, 2013

The Gap-Year Guidebook 2018
- www.gap-year.com, John Catt Educational Ltd, 2017

The Rough Guide to First-Time Around The World
- Doug Lansky, Rough Guides, 2016

Travel the World Without Worries: an Inspirational Guide to Budget Travel
- Marek Bron, CreateSpace Independent Publishing Platform, 2015

Work Your Way Around the World
- Susan Griffith, Trotman, 2017

Best Ever Travel Tips: Get the Best Travel Secrets and Advice from the Experts (Lonely Planet)
- Lonely Planet, 2014

The Backpacker's Bible
- Suzanne King, Portico, 2010

Volunteer: A Traveller's Guide to Making a Difference Around the World
- Lonely Planet, 2017

Archaeo-Volunteers: The World Guide to Archaeological and Heritage Volunteering
- Vacation Work 2009

# Improve Your Grades - Revision Courses

The following independent colleges offer retake, revision and university preparation courses, giving you the opportunity to improve your performance in a wide range of subjects, often during the Easter vacation.

## Abbey DLD Colleges

Abbey DLD is a group of independent sixth form colleges based in Cambridge, London and Manchester. The colleges offer a wide range of GCSE and A level courses, with an emphasis on rigorous but informal teaching, small class sizes, frequent testing and expert teachers.

**Twitter:** @abbeycolleges

**Web:** www.abbeycolleges.co.uk

**Facebook:** www.facebook.com/DLDcollege

## Ashbourne Sixth Form College, London

Easter revision courses seek to motivate students to strive for the highest grades at A level and GCSE, and to develop independence and self-reliance. As a private college, Ashbourne is able to offer classes which average five students per group.

Twitter: @AshbourneLondon

Web: www.ashbournecollege.co.uk

Facebook: www.facebook.com/ashbournecollege

## Bath Academy

Revision at Bath Academy features small group tuition and bespoke one-to-one tuition in all major A level and GCSE subjects. Sessions are split between formal teaching focused on areas that students find challenging and daily assessment to improve exam technique and application of knowledge. Detailed feedback is given on assessments and an end of course report is sent. Accommodation is available.

Twitter: @BathAcademy

Web: www.bathacademy.co.uk

Facebook: www.facebook.com/Bath.Academy

## Cambridge Centre for Sixth-form Studies

All main GCSE and A level subjects are offered for Easter revision, taught in small groups by qualified, experienced subject specialists, skilled in managing directed revision programmes. Courses cover intensive revision of core topics and exam practice. Boarding accommodation is available.

Twitter: @CCSSCambridge

Web: www.ccss.co.uk

Facebook: www.facebook.com/CCSS.SixthForm

### Carfax College, Oxford

Easter revision is available in all subjects offered at A level (AS/A2) and GCSE. Subject tutorials are taught individually (one-to-one), while study skills and exam technique are taught in small groups. Formal mock exam practice is included, and you can attend morning and/or afternoon sessions.

Twitter: @CarfaxEducation

Web: www.carfax-oxford.com

Facebook: www.facebook.com/carfaxoxford

### Chelsea Independent College, London

All main subjects are offered for AS, A2 and GCSE revision, while IELTS preparation for international students is also available. You can take 20- or 40-hour sessions, featuring intensive tuition in small classes, comprehensive revision of the exam syllabus, development of key skills and knowledge, focus on exam technique and frequent exam practice on past papers. Boarding accommodation is available.

Twitter: @ChelseaCollege

Web: www.chelseaindependentcollege.com

Facebook: www.facebook.com/ChelseaIndependentColl

### Collingham College, Kensington, London

Intensive tuition in small classes enables individual needs to be met. Groups are Board-specific as appropriate and are formed according to syllabus, topics, texts and so on. The courses are planned to give a clear understanding of the essentials of the syllabus and to teach exam techniques, so that you can use your knowledge to best effect.

Twitter: @CollinghamLdn

Web: www.collingham.co.uk

Facebook: www.facebook.com/CollinghamCollege

## Duff Miller Sixth Form College, South Kensington, London

Duff Miller offers Easter Revision courses across a wide range of A level and GCSE subjects. The College aims to inspire, motivate and help you fulfil your academic potential. The primary emphasis of the revision courses is to enhance subject knowledge and establish a rigorous, disciplined and effective approach, resulting in peak exam performance.

Twitter: @DuffSixthForm

Web: www.duffmiller.com

Facebook: www.facebook.com/DuffMillerSixthFormCollege

## Greene's Tutorial College

Oxford's oldest tutorial college, Greene's has been running highly successful one-year, retake and short revision GCSE and A level courses for 50 years. In addition to timely and well-focused advice on examination revision for all ages and levels, Greene's offers very effective intensive courses for the multitude of general and subject-specific aptitude tests required for many of the competitive degrees and top universities in the UK or the US.

Twitter: @GreenesNews

Facebook: www.facebook.com/greenes.org.uk

Website: www.greenes.org.uk

## Harrow School Enterprises

Harrow School Easter Revision Courses are suitable for students taking GCSE, IGCSE, AS and A2 exams. They are designed to improve grades by increasing your knowledge, refining your examination technique and boosting your confidence. Attending Harrow's revision courses in the Easter holidays can make the difference between a B and an A grade or get you a place at your first choice university rather than your second.

Twitter: @harrowschoolent

Web: www.harrowschoolenterprises.com

Facebook: www.facebook.com/HarrowSchoolEnt

### HillCrest Revision

HillCrest run intensive examination revision courses for GCSE, IGCSE, A level, IB and SAT students during holidays. Courses are based at centres in London, Brighton, Canterbury, Mickleham, Birmingham, Manchester, Malaysia and Dubai.

Web: www.hillcrestrevision.com

Facebook: www.facebook.com/HillCrestRevisionCourses

### Justin Craig Education

Established in 1981 and with 21 centres around the country, Justin Craig Education provides GCSE, AS and A2 group revision courses. Offering small, informal tutorial groups taught by friendly, enthusiastic and experienced tutors, Justin Craig Education has helped over 100,000 students achieve the grades needed to get into their chosen universities.

Twitter: @JustinCraig_

Web: www.justincraig.ac.uk

Facebook: www.facebook.com/justincraigrevision

### Lansdowne College, London

Lansdowne College has run Easter Revision courses for over 25 years, helping students to realise their full academic potential at both GCSE and A level, and enabling them to progress onto their chosen path for their studies. Students are often unaware of exactly what examiners are looking for and subsequently unsure of what makes an A grade answer. The Easter Revision courses remedy this by focusing not just on the subject content, but also on extensive and comprehensive examination preparation, including a mock examination at the end of the course.

Twitter: @lansdownecoll

Web: www.lansdownecollege.com

Facebook: www.facebook.com/LansdowneColl

### Mander Portman Woodward (MPW)

MPW is an independent sixth form college group, with colleges in London, Birmingham and Cambridge. In addition to full-time GCSE courses, AS courses and A2 courses over a very wide range of subjects and with no restrictions on subject combinations, MPW offers intensive A level and GCSE retake courses and revision courses over Easter. Characterising all courses is an absolute maximum of eight students within any one class and a strong emphasis on exam technique and exam practice.

Twitter: @MPWLondon; @MPWCambridge; @MPWBirmingham

Web: www.mpw.ac.uk

Facebook: www.facebook.com/Mander-Portman-Woodward-college-United-Kingdom-156064537793950/

### Millfield School, Somerset

On its Easter Revision course, Millfield offers specialist tuition at GCSE, AS and A2 level, in a broad range of subjects to small groups of students, providing a balance of taught content and rehearsal of technique. Students from all schools are welcome. Courses are offered on both a residential and non-residential basis.

Twitter: @MillfieldEnts

Web: www.millfieldenterprises.com

Facebook: www.facebook.com/MillfieldEnterprises

### Oxford Tutorial College

Oxford Tutorial College organises short intensive revision courses at Easter, both at A level and GCSE, as well as supplementary teaching at other times throughout the year to support the work being done at school. There is a mixture of individual tuition and small group seminars.

Web: www.oxfordtutorialcollege.com

Facebook: www.facebook.com/Oxfordtutorialcollege

## Revision Courses Reigate

Revision Courses have been in operation for over 30 years at Reigate College in Surrey, offering structured day courses for GCSE, IGCSE, AS and A2 students. Tutors customise each course to the needs of students. Therefore, whilst the courses for each subject cover the core material, areas of specific difficulty can be incorporated into the schedule.

Web: www.therevisioncourses.co.uk

## Rochester Independent College, Kent

Intensive Easter revision courses at Rochester provide an opportunity to gain an overview of the syllabus and a chance to practise applying knowledge to real examination questions. This helps with recall of facts and hones skills required for accurate question interpretation and structuring full but concise answers. Most importantly, the courses give a real confidence boost at a crucial time.

Twitter: @rochesterindcol

Web: www.rochester-college.org

Facebook: www.facebook.com/RochesterIndependentCollege

# Taster Courses

# Taster Courses

Designed mainly for Year 12 students, taster courses give you the opportunity to find out about subjects you could study and the careers to which they might lead. You may have the chance to explore a particular subject in depth and to taste what it is like to study at a particular university.

You should note that taster courses are not the same as pre-application Open Days, which are organised by almost all universities and colleges. We focus here on courses designed to give you a more detailed opportunity to experience academic and social life on campus. You should also bear in mind that, while we refer to Taster Courses for the sake of clarity and simplicity, universities are independent institutions and are free to use whatever terminology they like. It is, therefore, not unusual to find Taster Courses described as Summer Schools or Summer Academies (at any time of year!), Campus Days or Find Out More Days.

## Duration

Course lengths vary: some are one-day courses only, others may last a weekend or even a week. You may have the chance to stay overnight on campus. Most courses are free, although you will normally have to pay for travel and accommodation.

## Format

Most Taster Courses include lectures, discussions and tutorial sessions, so that you can meet the departmental staff and get hands-on experience of using the facilities. This can provide an important insight into how the university or college operates. Taster Courses should also allow you to find out about other aspects of undergraduate life, such as sporting, musical, drama and cultural activities, accommodation and other amenities.

## Benefits of Attending a Taster Course

- You are more likely to choose a higher education course to suit your interests and abilities
- You can highlight your attendance in the personal statement section of your UCAS application
- You may find that a course or campus is not for you, and decide to reject it in favour of others
- You can discuss your impressions during interviews with university or college admissions tutors

## Other Taster Courses in this Guide

In addition to the Taster Courses listed on the following pages, see our section on Academic, Cultural, Scientific, Technological, Teaching and Sports-related options for details of other taster courses, particularly those in Art, Design, Fashion and Performance, Medicine and Veterinary Science, and Science, Technology, Engineering and Maths. There are also options under the Summer Schools heading (see page 99).

## Aston University, Birmingham

Aston organises a range of Masterclasses and Sixth Form Conferences providing a taster of university life in specific subject areas such as Biology, Business, Chemistry, English Language, Engineering, Law, Mathematics, Modern Foreign Languages (French, German and Spanish), Optometry, Pharmacy, Psychology, Politics and Sociology. Masterclasses are free to attend but booking is essential. They offer the opportunity for Year 12 and 13 students both to enrich your current studies and to discover more about studying a particular subject at degree level. Programmes for Masterclasses and Sixth Form Conferences are announced from September. There are also residential events announced in September.

Further information can be obtained by emailing:
sro@aston.ac.uk

or by visiting the website at:

www.aston.ac.uk/study/undergraduate/sro/pre-16/on-campus

**Twitter:** @AstonUniversity

**Facebook:** www.facebook.com/astonuniversity

## University of Bath

Bath organises Year 12 Taster Days with schools and colleges who bring groups of students to the campus. Schools are responsible for arranging and paying for transport to the University. Students can also apply to attend individually.

Subject Taster Days are designed to give you a full experience of the University and include opportunities for a tour of the campus, a chance to meet and talk to current University students as well as subject specific sessions with University academic staff.

Full details can be found on the website at:

www.bath.ac.uk/campaigns/bring-your-year-12-students-to-an-a-level-enrichment-day

**Twitter:** @UniofBath

**Facebook:** www.facebook.com/uniofbath

## University of Birmingham – Physics

Successful and popular residential summer schools have taken place for the past 12 years in Birmingham's School of Physics and Astronomy. The schools are designed to give AS students studying Physics and Mathematics a taste of lectures, problem solving classes and laboratory work at university, as well as offering experience of an overnight stay and a chance to visit Birmingham city centre.

Full details at:

www.birmingham.ac.uk/schools/physics/outreach/Secondary-Schools/summer-schools.aspx

**Twitter:** @unibirmingham

**Facebook:** www.facebook.com/unibirmingham

## University of Central Lancashire (Uclan)

To help you make important decisions about the future, Uclan invites Year 12 learners to a programme of University Experience Days. In each event, professional advice is shared on important student issues such as the UCAS application process and challenges to expect when moving to university from school/college. Each day covers a theme such as Law, Languages, Psychology, Social Sciences, Civil Engineering, Business and Management or Health, with talks delivered by expert, research-active academics.

Full details at:

www.uclan.ac.uk/visit/university_experience_days.php

**Twitter:** @UCLan

**Facebook:** www.facebook.com/officialUCLan

## Cornwall College Newquay

Visitors from all parts of the UK can attend a university taster day at Cornwall College Newquay, to find out more about the zoology, surf and marine courses on offer at the specialist coastal campus. The College offers foundation and BSc degrees in subjects that are rarely found elsewhere. During the day, you can learn more about the courses, the annual field trips to such places as Borneo, Egypt and South Africa, and related work experience opportunities. You would also tour the Newquay campus, visit Newquay Zoo, the Blue Reef Aquarium or Tolcarne beach, take part in a team building task and experience an admissions interview.

To book your place, call 01637 857957,

email: **newquay.enquiries@cornwall.ac.uk**

or visit the website at:
www.collegeexperience.co.uk

**Twitter:** @CornwallCollege

**Facebook:** www.facebook.com/cornwallcollegeuk

## University of East Anglia (UEA)

UEA summer schools are for students in year 12 from all backgrounds. You must be aiming to achieve A and B grades in your A-levels (or equivalent). You can take part in hands-on subject tasters, explore campus living by staying in university accommodation and meeting current students. Subjects already confirmed for 2017 include: Chemistry and Pharmacy, Engineering, Geography, Law and Mathematics.

Full details from:

www.uea.ac.uk/study/info-for/young-people/post-16/university-tasters-enrichment/summer-schools

**Twitter:** @uniofeastanglia

**Facebook:** www.facebook.com/ueaofficial

## Edge Hill University

Edge Hill offers opportunities for Year 12 (Year 13 if in Northern Ireland) to experience life on campus during the course of a four-day, three-night summer residential visit. A wide range of departments takes part and you can select the subject area you are most interested in pursuing at university, in order to get a real insight into what it would be like to study that subject. All attendees have access to the University library and IT network for the duration of the residential. Evening activities (some compulsory and some optional) run each night, and you stay in halls of residence. Current Edge Hill University undergraduate students are on hand throughout the residential, offering support and guidance where necessary.

Full details from: www.edgehill.ac.uk/studentrecruitment/year-12-residential

**Twitter:** **@edgehill**

**Facebook:** **www.facebook.com/edgehilluniversity**

## University of Exeter - Exeter Scholars

Exeter Scholars is designed to encourage disadvantaged young people to find out more about university though a range of different activities and opportunities. Your experience as an Exeter Scholar participant will include a range of activities and opportunities appropriate for your year group and place of living, and will provide you with an insight into university as well as a subject specific experience. Year 12 students are invited to attend a residential event and alongside your chosen academic programme, the residential experience will also give you the chance to explore the local area, develop your cooking skills, try out study societies, and find out what student life might be like on a University of Exeter campus.

For further details visit the website at: **www.exeter.ac.uk/exeterscholars**

**Twitter:** **@UniofExeter**

**Facebook:** **www.facebook.com/exeteruni**

## University of Huddersfield

Taster Days at Huddersfield give you the chance to experience first-hand what it is like to study a subject at university. On the day, you take part in a sample lecture and interactive workshops. Your visit may also include a tour of the campus, together with an opportunity to meet current students and find out all you want to know about student life. You will develop a practical understanding of a subject area, learn about the range of courses available within that subject area and make informed decisions about your future.

Full details from: www.hud.ac.uk/subject-specific-activities

**Twitter:** **@HuddersfieldUni**

**Facebook:** **www.facebook.com/huddersfielduniversity**

## Lancaster University

Lancaster's summer school programme offers four-day residential courses specialising in: Medicine; Biochemistry and Biomedical Science; Arts, Humanities and Social Science; Business and Management; Science and Technology.

For further information, call 01524 595013,

email: outreach@lancaster.ac.uk

or visit the website at: www.lancaster.ac.uk/summer-schools

**Twitter:** **@LancasterUni**

**Facebook:** **www.facebook.com/lancasteruniversity**

## Liverpool John Moores University (LJMU)

The LJMU Summer University offers free three-day residential programmes in the Built Environment, Business, Computing, Engineering and Mathematics. If you are interested in fashion and design, you might consider the LJMU Fashion Project, which involves attendance once a week after school, over an eight-week period, at the John Lennon Art and Design Academy.

For further information, call 0151 904 6385, email outreach@ljmu.ac.uk or visit the website at: www.ljmu.ac.uk/study/undergraduate-students/services-for-schools-colleges-and-prospective-students/events-and-opportunities

Twitter: @LJMU

Facebook: www.facebook.com/LJMUofficial

## London University Taster Course Programme

With over 300 courses in more than 100 different subject areas at numerous university institutions in and around London, the University Taster Course Programme provides the opportunity to experience life as a university student in the subject area of your choice, ranging from accounting, anthropology or astrophysics to veterinary science, visual culture or war studies. Through your chosen taster course(s) you will experience the different teaching methods used by university academics and gain an insight into the additional facilities available, whilst meeting students from across the UK. The taster courses run for from half a day to one week, and are available throughout the year with the majority taking place between March and July. All courses are non-residential and are provided free of charge.

Visit the website at: www.london.ac.uk/tasters or email: tastercourses@london.ac.uk for full details of the 2019 programme.

In 2018 the following universities and colleges participated:

| Institution | Subject |
|---|---|
| Birkbeck | Accounting |
| | Arts and Humanities |
| | Business |
| | Computer Science |
| | Creative Writing |
| | Cultures and Languages |
| | Economics |
| | English |

| | |
|---|---|
| | Film and Media |
| | French |
| | German |
| | Global Cinemas and Screen Arts |
| | History of Art |
| | History of Art with Curating |
| | History of Art with Film |
| | History of Art and History |
| | Information Systems |
| | Japanese |
| | Journalism and Media |
| | Management |
| | Marketing |
| | Mathematics |
| | Media and Culture |
| | Organizational Psychology |
| | Portuguese |
| | Psychology |
| | Psychology for Education |
| | Spanish |
| | Statistics |
| | Theatre and Drama |
| City University | Accounting |
| | Banking and International Finance |
| | Business and Management |
| | Civil Engineering |
| | Computer Science |
| | Criminology |
| | Economics |
| | Electrical and Electronic Engineering |
| | Engineering |
| | English |
| | Finance |
| | Health and Social Care |
| | History |
| | International Politics |
| | Investment and Financial Risk Management |
| | Journalism |
| | Law |
| | Mathematics |
| | Mechanical and Aeronautics Engineering |
| | Media and Communications |
| | Midwifery |
| | Music |
| | Nursing |
| | Optometry |
| | Politics |
| | Psychology |
| | Radiography |
| | Sociology |
| | Speech and Language Therapy |

| | |
|---|---|
| | Structural Engineering |
| Courtauld Institute of Art | History of Art |
| Goldsmiths | Anthropology<br>American Literature<br>Art and Design<br>Arts Management<br>Business and Management<br>Child Psychology<br>Classics<br>Communications<br>Computing<br>Creative Computing<br>Criminology<br>Curating<br>Design<br>Drama<br>Economics<br>Education Culture and Society<br>Educational Studies<br>English and Comparative Literature<br>English and History<br>English Literature<br>Entrepreneurship<br>Film Studies<br>Fine Art<br>Geography<br>Graphic Design<br>Health and Social Care<br>History<br>History of Art<br>International Relations<br>Journalism<br>Illustration<br>Mandarin for Beginners<br>Marketing<br>Media Studies<br>Music<br>Musical Theatre<br>Philosophy<br>Photography<br>Politics<br>Psychology<br>Religious Studies<br>Social Work<br>Sociology<br>Theatre and Performance<br>Therapeutic Studies<br>Visual Cultures |
| Imperial College | Applied Mathematics<br>Computer Science |

| | |
|---|---|
| | Computing<br>Mathematics |
| King's College | Archaeology<br>Biomedical Engineering<br>Business and Management<br>Catalan<br>Chemistry<br>Classics<br>Classical Studies with English<br>Comparative Literature<br>Computer Science<br>Digital Culture<br>Electronic Engineering<br>English Language<br>English Literature<br>European and International Studies<br>Film Studies<br>French<br>Geography<br>German<br>Global Affairs<br>Global Health and Social Medicine<br>History<br>International Development<br>Liberal Arts<br>Linguistics<br>Mathematics<br>Midwifery<br>Music<br>Nursing (Adult Nursing)<br>Nursing (Children's Nursing)<br>Nursing (Mental Health Nursing)<br>Nutrition<br>Philosophy<br>Physics<br>Political Economy<br>Politics<br>Portuguese<br>Psychology<br>Religious Studies<br>Spanish<br>Theology<br>War Studies and International Relations |
| LSE | Anthropology<br>Economic History<br>Finance<br>Geography and Environment<br>International History<br>Philosophy<br>Social Policy |

| | |
|---|---|
| Middlesex University | Architectural Technology<br>Mathematics<br>Physical Computing and Robotics |
| Pearson College London | Accounting<br>Animation<br>Business<br>Game Art<br>Law<br>VFX<br>Visual Effects |
| Queen Mary | Astrophysics<br>Computer Science<br>Computer Science with Business Management<br>Creative Computing<br>Dentistry<br>Drama<br>Economics<br>Electronics<br>Engineering<br>Engineering (Aerospace)<br>Engineering (Biomedical)<br>Engineering (Chemical Engineering)<br>Engineering (Design, Innovation and Creative)<br>Engineering (Materials)<br>Engineering (Robotics Engineering)<br>English and Drama<br>Environmental Science<br>Geography<br>Global Health<br>Information Technology with Management for Business (ITMB)<br>Material Science<br>Mathematics<br>Mathematics with Business<br>Mathematics with Finance<br>Medicine<br>Physics<br>Software Engineering for Business |
| Royal Holloway | American Literature<br>Biosciences<br>Classics<br>Comparative Literature<br>Computer Science<br>Criminology and Psychology<br>Criminology and Sociology<br>Digital Media and Culture<br>Economics and Management<br>English Literature<br>Film Studies |

| | |
|---|---|
| | Film and TV Production<br>History<br>International Relations<br>Law<br>Media Arts<br>Physics<br>Physics (Theoretical Physics)<br>Politics<br>Psychology<br>Sociology |
| Royal Veterinary College | Animal Behaviour, Welfare and Ethics<br>Biological Sciences<br>Veterinary Medicine<br>Veterinary Nursing |
| School of Oriental and African Studies | Africa Studies<br>Archaeology<br>China and Inner Asia Studies<br>Development Studies<br>Economics<br>Finance and Management<br>History<br>History of Art<br>International Development<br>Japan Studies<br>Korea Studies<br>Linguistics<br>Near and Middle East Studies<br>Politics and International Studies<br>Religions and Philosophies Studies |
| St George's | Medicine |
| University College | Anthropology<br>Biochemical Engineering<br>Classics<br>Earth Sciences<br>East European Languages<br>Education Studies<br>Mathematics<br>Medical Physics<br>Natural Sciences<br>Pharmacology<br>Population Health<br>Psychology with Education<br>Social Sciences<br>Social Science with Quantitative Methods<br>Urban Planning |

| | |
|---|---|
| University of East London | Architecture<br>Computing<br>Engineering<br>Healthcare and Biomedical Research<br>Psychology<br>Social Sciences |
| University of Greenwich | Architecture and Landscape<br>Biology<br>Built Environment<br>Chemistry<br>Computing and Information Systems<br>Computer Science<br>Computer Security and Forensics<br>Computer Systems and Networking<br>Creative Writing and Drama<br>Creative Professions<br>Design<br>Digital Arts<br>Engineering (Chemical Engineering)<br>Engineering (Civil Engineering)<br>Engineering (Electrical Engineering)<br>Engineering (Mechanical Engineering)<br>Engineering (Design, Product Engineering)<br>English Literature<br>English Language<br>Forensic Science<br>Geography and Environmental<br>History, Political and Social Sciences<br>Law and Criminology<br>Mathematics<br>Mobile Applications Development<br>Psychology<br>Software Engineering<br>Sports Science<br>Theatre |
| University of London Institute in Paris | French |
| University of Westminster | Biology |

**Twitter: @LondonU**

**Facebook: www.facebook.com/LondonU**

## London Metropolitan University

Free summer schools give you the chance to explore university life and discover what your course will really be like. You'll attend lectures, seminars and hands-on workshops and get tips on your UCAS application and study skills. You'll also learn first-hand from student ambassadors what student life is like. Faculties taking part include: Life Sciences and Computing; Social Sciences and Humanities; Business and Law; Art, Architecture and Design.

For further information, email summerschool@londonmet.ac.uk or visit the website at: www.londonmet.ac.uk/about/school-programmes-and-outreach/summer-school

**Twitter:** @LondonMetUni

**Facebook:** www.facebook.com/londonmetuni

## Loughborough University Engineering Experience

Loughborough's annual, two-day event for Year 12 students gives an insight into what it is like to study engineering at university. It's your chance to experience first-hand the Loughborough campus and student accommodation. You'll also have an opportunity to take part in a number of different taster events within the engineering departments.

Staff and students are on-hand throughout the two days to talk about the courses on offer and answer other questions you may have. Department experiences include: Aeronautical and Automotive; Chemical; Civil and Building; Electronic, Electrical and Systems; Manufacturing; Materials; Mechanical; and Product Design Engineering. There is a small charge for this event to cover meals and accommodation.

For further information, email: engexp@lboro.ac.uk or visit the website at:

www.lboro.ac.uk/study/school-college-liaison/events/engexp

## University of Manchester

Manchester's Discover Days are informative, interactive events for aspiring undergraduate students who are interested in certain subject areas but may not know exactly which course to choose. They are ideal for Year 12 students, or for other people applying to university. Discover Days give first-hand experience of university study and student life, and help you make informed decisions about your university choices. Subjects available in 2016 are: Sociology, English, History, Chemistry, Psychology, Life Sciences, Philosophy, Computer Science, Law, Midwifery, Nursing, Engineering.

Full details from: www.manchester.ac.uk/study/undergraduate/aspiring-students/discover-days

**Twitter:** @OfficialUoM

**Facebook:** www.facebook.com/TheUniversityOfManchester

## University of Reading Food and Nutritional Sciences Summer School

If you are a young scientist thinking about a career in food, the Department of Food and Nutritional Sciences at Reading offers an annual short introductory course in June/July for lower 6th form students (Year 12). The course is sponsored by manufacturing and retail companies in the food industry.

The degree courses in these subjects are multidisciplinary, encompassing chemistry, microbiology, nutrition, quality assurance, sensory analysis, management, food engineering and manufacturing. They are very practical and frequently involve industrial training.

For further information, email: food@reading.ac.uk or visit the website at:

www.reading.ac.uk/food/shortcourses/foodbio-shortcourses.aspx

## Reading Real Estate Foundation - Pathways to Property

The Pathways to Property initiative is a programme for Year 12 students in UK state schools and colleges who would like to find out more about a career in the property sector. Led by the Reading Real Estate Foundation (RREF) at the University of Reading, the initiative is supported by British Land, one of the largest property development and investment companies in the UK, the Sutton Trust and other leading real estate firms and charitable trusts. Activities include a free Summer School at the University of Reading, Work Shadowing and a series of industry-led talks.

For full details, visit the website at:

rref.henley.ac.uk/rref-what-we-do/Pathway_to_Property/rref_pathways_to_property_summer_school_2018.aspx

**Twitter:** **@PathstoProperty**

**Facebook:** **www.facebook.com/PathwaysToProperty**

## Royal Agricultural University (RAU)

The RAU offers a two-day residential taster course for 16- to 18-year-olds designed to give the opportunity to explore the campus and get a taste of what it's like to live and study there. You will stay in student accommodation, take part in classroom lessons and fun practical activities, and get lots of information and advice about coming to university. You can choose whether you would like to participate in the 'surveying' or 'management' taster sessions, with 'surveying' covering the Real Estate and Land Management courses, and 'management' incorporating Agriculture, Food, Environment, Equine and Business.

For more information email: opendays@rau.ac.uk, call 01285 652531 or visit the website at:

www.rau.ac.uk/study/undergraduate/open-days/discover-rau

**Twitter:** **@RoyalAgUni**

**Facebook:** **www.facebook.com/RoyalAgUni**

## Sparsholt College Hampshire

If you are considering a career in the land, environment or animal industries, Sparsholt College Hampshire offers a number of taster days. Taster days are run during the school holidays for Year 10 and 11 students to experience a real day at Sparsholt. Subjects include Agriculture, Agricultural Engineering, Animal Management, Arboriculture and Forestry, Equine Studies, Fishery Studies, Gamekeeping, Horticulture, Motor Vehicle and Engineering, Sport, Outdoor Education and Public Services.

Full details are available from the website at:

www.sparsholt.ac.uk/taster-days

**Twitter:** **@Sparsholt_Coll**

**Facebook:** **www.facebook.com/SparsholtCollege**

## Swansea University Engineering Summer School

Swansea's College of Engineering runs a Summer School every year for year 12 Students. The residential course is designed to give an insight into all the engineering courses on offer and to help you decide whether an engineering degree is likely to be for you.

Further information from the website at:

www.swansea.ac.uk/engineering/summer-schools

**Twitter:** **@SUEngineering**

**Facebook:** **www.facebook.com/SwanseaUniEngineering**

## Teesside University

Taking part in Teesside's year 12 residential summer school means that you have the chance to participate in academic sessions covering a range of subjects that you could study if you were a university student. You will interact with current students and learn how to be a successful student yourself. You will spend four days and three nights on campus, staying in the Parkside halls of residence.

Full details from: www.tees.ac.uk/summerschool

**Twitter:** **@TeessideUni**

**Facebook:** **www.facebook.com/TeessideUni**

## University of York

York's Taster Days give Year 12 students the opportunity to find out about subjects you could study and the careers to which they might lead. You take part in three mini-lectures, meet staff and students, hear about student life and tour the campus. Taster Days take place in March each year, and online bookings open in early spring. Subject Conferences are similar but give you the chance to explore a particular subject in depth and have a taste of what it is like to study at York. The day includes lectures, practical workshops, a campus tour and the opportunity to talk to staff and current students.

For full details of taster days and subject conferences, visit the website at:

www.york.ac.uk/study/schools-and-colleges/post-16/subject-conferences

**Twitter:** **@UniofYork**

**Facebook:** **www.facebook.com/universityofyork**

# Indexes

## Taster Course Index

# Organisation Index

# Geographical Index

# Subject Index